Yale Agrarian Studies Series
JAMES C. SCOTT, *series editor*

The Agrarian Studies Series at Yale University Press seeks to publish outstanding and original interdisciplinary work on agriculture and rural society—for any period, in any location. Works of daring that question existing paradigms and fill abstract categories with the lived experience of rural people are especially encouraged.
—JAMES C. SCOTT, *Series Editor*

For a complete list of titles in the Yale Agrarian Studies Series, visit www.yalebooks.com/yupbooks/seriespage.asp?series=94.

Remoteness and Modernity

Transformation and Continuity
in Northern Pakistan

Shafqat Hussain

Yale UNIVERSITY PRESS
NEW HAVEN & LONDON

Published with assistance from the Mary Cady Tew Memorial Fund.

Yale University Press books may be purchased in quantity for educational, business, or promotional use. For information, please e-mail sales.press@yale.edu (U.S. office) or sales@yaleup.co.uk (U.K. office).

Set in Ehrhardt type by IDS Info Tech Ltd., Chandigarh, India.
Printed in the United States of America.

ISBN 978-0-300-20555-8

Catalogue records for this book are available from the Library of Congress and the British Library.

This paper meets the requirements of ANSI/NISO Z39.48–1992 (Permanence of Paper).

10 9 8 7 6 5 4 3 2 1

FOR
Valida, Annie, and Khadija

Contents

Acknowledgments

I would not have been able to write this book without the moral, social, and intellectual support and love of Annie, my wife. Her belief in my ability to complete a PhD was instrumental in sustaining my own determination and persistence. My children, Khadija, Musa, and Askari, give me inspiration and joy in life.

I would also like to thank Michael Dove, who provided overall guidance on this project. I could not have had a better instructor. His comments on my work were not only of instrumental value but affected the way I thought about history and social analysis. I am not sure if I am more impressed with his immense theoretical and empirical knowledge or his wisdom. K. Sivaramakrishnan enabled me to place my work in wider theoretical and re-gional traditions. I particularly benefited from Shivi's almost encyclopedic knowledge and his detailed and remarkably insightful comments that always urged me to push my work to the next level of sophistication. His remarks helped me to fine-tune and streamline my work and to make my arguments clear and concise. Carol Carpenter was a source of constant intellectual advice and encouragement; I particularly benefited from her advice on the symbolic analysis of the Shimshalis' economic and social rituals.

I am thankful to Jim Scott, whose initial comments on the first four chapters set the intellectual path I take in this book. He encouraged me to think through spatial categories of power and resistance and their transfor-mation through time. I am also thankful for his encouragement to submit the book to the Yale Agrarian Studies Series. I would also like to thank

Elisabeth Barsa in the Doctoral Office of the Yale School of Forestry and Environmental Studies for providing administrative support during the course of my doctoral study. Elisabeth always answered my questions with a smiling face and helped me with numerous mundane matters. I must also thank the two reviewers who provided feedback on an earlier draft of the manuscript. I am particularly indebted to the first reviewer, whose comments inspired me to reframe my argument in terms of remoteness rather than marginality.

My greatest thanks must go to the people of Shimshal who hosted me and provided me with the knowledge and information that form the core of this book. I am particularly thankful to Muzaffar-ud-Din, who encouraged my work in Shimshal and facilitated my research at every step with his generous offers of time and acts of hospitality. Inayat Ali helped me understand the conflict over Khunjerab National Park in its historical context and provided me with invaluable unpublished documents on the issue. Master Daulat Amin and Master Sher Ali provided me with the history of Shimshal and helped me understand the meaning and significance of Shimshali rituals and festivals. In this book, all names of the people of Shimshal have been changed. In Passu, the owner of Passu Inn, Ghulam Mohammad, helped me understand the modern history of the Hunza state within the context of wider Pakistani politics. I am especially thankful to Mr. Ghulam Mohammad for the hospitality he extended whenever I stayed at Passu Inn. I am grateful to Colonel (retired) Saad Ullah Beg, a scion of the wazir family in Hunza, who helped me understand the history of Hunza state from a new perspective. Mr. Sher Baz Burcha of Gilgit Municipal Library helped me navigate various historical sources and documents. Although in this book I criticize George Schaller's work, I respect him enormously as a naturalist and I am grateful for the guidance and support he has given to me personally.

During my fieldwork I was often accompanied by Ghulam Mehdi, Shariat Hussain, and Khawar Javed Bhatti, who were excellent company. I would especially like to thank them for offering an alternative point of view to mine and listening to my endless theorizing and countertheorizing. In Skardu, the staff of AKRSP provided me logistical support throughout the research period. I am particularly thankful to Mr. Jawad Ali, who provided not only administrative and logistical help but also intellectual guidance. I am also indebted to Brigadier Yousaf Majoka of the Pakistan Army, who helped me through the many security hurdles I encountered during my research. In New Haven, the Yale School of Forestry and Environmental

Studies' Dove-Carpenter doctoral lab provided an extraordinary forum for discussing ideas, and the participants gave me excellent feedback while I was writing the first draft of the book. I am particularly indebted to Jonathan Padwe, Andrew Mathews, Ann Rademacher, David Kneas, Julie Velasquezrunk, Christiana Ehringhaus, and Steve Rhee. I also had the good fortune to be actively involved with the Yale Agrarian Studies Colloquium series. The presentations and discussions each week were inspiring. I want to thank Chris Moody, Todd Holmes, Eric Rutkow, and Vikram Thakur of New Haven for helping me through moments when the ideas refused to flow. Special thanks to Cathy Shufro, who gave critical advice in making the first three chapters coherent and clear. Chris Milan produced the maps, and I appreciate his patience in dealing with my numerous requests for amendments and additions. At Yale University Press, Samantha Ostrowski helped format the manuscript and helped me navigate the administrative process. Jean Thomson Black, the senior executive editor at Yale University Press, was instrumental in guiding me through the publishing process and advising me on the selection of images and titles. My copy editor, Robin DuBlanc, was absolutely brilliant in her editorial work. Her suggestions improved the quality of the manuscript tremendously.

Mirs of Hunza

1790?–1823   Salim Khan II
1823–63   Ghazanfar Khan I
1863–86   Ghazan Khan I
1886–91   Safdar Ali Khan
1892–1938   Nazeem Khan
1938–46   Ghazan Khan II
1946–74   Jamal Khan
1974   Hunza state abolished, but Mir Ghazanfar Khan II
             acts as the ceremonial mir

Chronology of the Hunza State and Its Relationships with Surrounding Polities

15th century	Hunza state founded; enters into tributary relationship with China
1842	Occupation of Gilgit by dogra of Kashmir
1846	State of Kashmir founded, its territorial extent demarcated at east of Indus
1847	First British emissaries sent to Hunza never arrive
1860	First unsuccessful invasion of Hunza by Kashmir forces
1865	Kashghar in Chinese Turkestan occupied by Kirghiz; Hunza suspends tribute to China
1866	Second unsuccessful invasion of Hunza by Kashmir forces
1869	Mir Ghazan enters into a treaty agreement with Kashmir
1872	Kashmir agrees to establishment of a British officer in Gilgit; British and Russians tentatively agree to mark the boundary of their relative spheres of influence at the source of the Oxus, just north of Hunza
1870–75	First and second mission of Douglas Forsyth to Kashghar and survey of Hunza passes by Biddulph
1878	Gilgit Agency established, Major John Biddulph as first political agent
1881	Gilgit Agency abandoned
1885	Panjdeh incident, Russian seizure of Afghan territory south of the Oxus
1886	Safdar Ali kills his father Mir Ghazan Khan to become ruler

1888	Colonel Gromchevsky visits Hunza
1889	Captain Francis Younghusband sent to Hunza to negotiate a deal with Safdar, Gilgit Agency reestablished
1891	British invade and conquer Hunza with the help of Kashmiri forces
1895	Pamir Boundary Commission demarcates Afghanistan's southern and northern boundaries to create buffer zone of Wakhan corridor between Russian and British empires
1895–1937	British Empire makes unsuccessful efforts to demarcate Hunza's boundary with China
1935	British Empire acquires Gilgit Agency on a sixty years' lease from Kashmir state
1937	Hunza stops paying tribute to China
1947	Hunza and others fight a "war of independence" against Kashmiri state forces and join Pakistan (Hunza remains a semiautonomous state)
1958	First motorized vehicle arrives in Hunza
1974	Prime Minister Bhutto abolishes Hunza state, merging it into Northern Areas
1978	Karakorum Highway completed
1988	Hunza and Northern Areas get a provisional legislative assembly
2009	Name of Northern Areas changed to Gilgit–Baltistan and Hunza becomes a separate district

Remoteness and Modernity

Introduction

L ate one afternoon in the autumn of 1889, a small party of
weather-beaten travelers descended into a cold, rugged valley
of the central Karakorum Mountains. The party was led by Francis
Younghusband, then a young captain in the British Indian army and one of
the greatest players of the Great Game, the nineteenth-century struggle
between the British and Russian empires for control of the tract of land
between South and Central Asia.[1] Coming over from Chinese Turkestan,
the party descended into the upper Braldu valley, part of the independent
princely state of Hunza that today lies within the Gilgit-Baltistan region of
northern Pakistan.[2] There the party encountered a group of men whom
Younghusband described as "wild looking . . . shouting and waving us back,
and pointing their matchlocks at us" (1896, 260). These men were from
Shimshal, the northernmost village of the Hunza state.[3] Located between
the fast-expanding Russian Empire to the north and British India to the
south, Hunza was still outside the influence of both; its head was a belliger-
ent ruler, the mir, who had succeeded his father that year by murdering him.
In the context of the Great Game, Younghusband's visit to this remote region was aimed at winning the allegiance of the mir and establishing a frontier against the southward-expanding Russian Empire. The Russian threat
never materialized, but within the next two years Hunza would become the
outermost post of the British Empire in India.

Almost one hundred years later, in 1986, another small party of weather-
beaten travelers arrived in Shimshal: European tourists. They had trekked

three days to the village from the Karakorum Highway, traversing narrow, winding gorges through a desolate mountain landscape before reaching the wide valley and green fields of Shimshal. One member of the party wrote in the guest book kept at the campsite at which they stayed: "Discovering Shimshal is discovering a lost paradise. The people of this valley show sincerity and openness not found in many places in the world. God be with them." This comment and similar entries portray Shimshalis as peace loving and unthreatening, living in harmony with one another and their environment amid threats from the outside forces of modernization. The tourists had come to this remote community of yak herders in hopes they might find relief from what they saw as the excesses and flaws of the modern world. When I started my fieldwork in 2003, this view of Hunza still prevailed among the tourists I encountered.

While much has changed, much has remained the same in Shimshal and Hunza between the visits of Younghusband and those of contemporary tourists. The journey is still grueling. Although the region is partly incorporated into the economy and bureaucracy of Pakistan, the authority of the state remains in many ways as incomplete and tenuous as that of the British Empire. Now, as in the nineteenth century, the people of Hunza—the Shimshalis in particular—are defined by their remoteness; and now, as then, they engage productively with the discourse of outsiders who represent them as remote. But what "remoteness" connotes has fundamentally changed. In the late nineteenth-century context of the Great Game, their remoteness meant they were considered savages and dangerous people, the "other," and in particular an adversarial other, as opposed to an alien other.[4] Today, their remoteness makes them a different sort of other, an "intimate other." Though the people of Hunza are still seen as geographically distant, off the grid economically and culturally, the defining characteristic of their remoteness is that they represent an ideal culture, living a way of life that is seen as threatened by modernization.

Edwin Ardener grappled with changing representation by outsiders in his study of the Bakweri of Mount Cameroon in West Africa:

> To the strange arrivals the village was either a scene of "traditional hospitality of a simple folk" or the location of incomprehensible reticence. The very act of having arrived was its own justification. Years later, the new arrivals were a unit of gendarmeries, for this was the remote area of all remote areas for the new Francophone government

and, like all areas of this peculiar type, not only perceived to be Shangri La but also the home of purported smugglers and spies. How shall the inhabitants of a "remote area" evaluate the arbitrary love-hate of its visitors? Are alternating periods of "unspoiledness" and violence their inevitable fate? After the destructions of one generation of strangers how is it that they are asked to play the role of an ideal society to the next, before being unthinkingly redeveloped or underdeveloped out of existence by the next? (1989, 215–16)

My intention in this book is to ask the same questions in the case of Hunza that Ardener asked of Cameroon. Over the course of a century, Hunza has been an independent princely state, then a district and a semi-autonomous state within the British Empire, later a part of the Pakistani-administered Gilgit Agency, and later yet an administrative district of the Gilgit-Baltistan region of Pakistan. During that hundred years, the people of Hunza have been variously represented by outsiders—including British colonialists, Pakistani state officials, and modern-day Westerners—as original Aryans, as slave traders and caravan raiders, as innocent primitives and healthy frontiersmen, as marginal citizens of the Pakistani state, as ideal hosts of global tourists, and as both indigenous conservationists and avaricious degraders of the environment. The one factor that remains constant is Hunza's status as a remote area. How the people of Hunza have responded to this "love-hate" relationship with outsiders and how they used their remote status to engage the outside world are themes that run throughout the book. In particular, I look at how some of the core ideas and practices associated with modernity and modernization since the late nineteenth century—exploration, boundary marking and delimiting territories, industrialization, nationalism, international tourism, and environmentalism—persistently engage with and produce the material and conceptual conditions of remote areas.[5]

This is, however, not only a study of modernity's and modernization's production of remote areas but also an analysis of how the inhabitants of remote areas participate as active agents in this production. I follow the general trajectory of anthropological theory since the 1960s, which has sought to dispel the notion that out-of-the-way places have remained isolated and unchanged, untouched by modernity (Wolf 1982; Wallerstein 1974). Most current work on tribes and rural people now takes for granted that such people have almost always maintained contact with the "outside" world in one form or another (Scott 2010; Dove 2011). Contemporary analyses of how remote communities are affected by modernization and indeed are themselves modern (Piot 1999) and cosmopolitan (Gidwani and Sivaramakrishnan 2003; Marsden 2008) have informed my work throughout.

Remote Areas and Other Spatial Concepts

The analytical category of remote areas belongs to a class of similar and well-studied socio-spatial concepts such as marginal areas, frontiers, borderlands, and wilderness.[6] The categories in this class have a common condition: they stand in contrast to the core cultural, economic, and political aspects of a society. The absence or presence of the state in these areas has a different logic than in other areas, although each of the categories is marked by state power to some degree. Culturally, the people of these regions are seen as the "other," but this alterity represents multiple constructions of the "other," each invested with different meaning. Despite their commonality, each of the categories in the class has a certain "semantic density" that distinguishes it from the others. Kirsten Hastrup explains that semantic density is related to "frequency, that is, to the question of which meanings will be implied more often than others when particular categories are invoked" (1989, 226). Each of these categories is defined and imagined in a slightly different way.

Marginal areas are generally defined as being out of the way (Tsing 1993) but not necessarily poorly connected, physically, to the center of political and cultural power. Sometimes these areas are geographically remote, but alternatively they may exist close to or even within centers of power: for example, slums on the outskirts of cities or urban ghettos. Marginal areas are often populated by minorities, groups that are disadvantaged and discriminated against, or people seen as criminal or backward (Tsing 1993; Li 2000; Cullen and Pretes 2000; Browdin 2003), though this often goes alongside their perception as sites of creativity from which sharp critiques of the dominant order of the society emerge. In comparison, remote areas are not generally defined in negative terms. For example, tourists seeking charm in their travels frame their search in terms of remoteness, not marginality.

Frontier areas are perhaps the most enduring and historically loaded category of those I have mentioned. They are seen as open zones in which central political powers exert influence and eventually bring them under control. Historically, they are thought of as locations highlighting struggle between civilization and barbarism (Curzon 1908). Thus their meaning is invested with alterity and otherness but with positive potential. Since frontiers are open zones with potential for expansion, there are no "real" frontiers left today. This has led to a slight alteration in the meaning of the category; frontier areas are now defined by their ambiguous regime of governance rather than their formal legal status. Many international resource-extraction corporations and entities work through establishing special access zones globally, where they create semi-legal regimes of management. Now described as zones rather than dividing lines (Lamb 1968), frontiers are also thought about in cultural terms. They represent creativity, but the creativity ascribed to people of frontier areas is different than that attributed to people of marginal areas. In frontier areas, creativity refers to the human ingenuity employed in the struggle between culture and nature; in marginal areas, creativity emerges out of asymmetrical power relationships between powerful and powerless groups.

Border areas are defined by their administrative and political conditions. Unlike frontiers, which are seen as open zones, borders mark clear delimitations of territory. Thus, they represent closure. The creativity of the people of border areas is defined by their negotiation of two different legal and political jurisdictions, two regimes existing side by side in sharp relief divided by a line of border demarcation. Border areas are unique among the categories we are discussing in that they represent state authority unambiguously; the state apparatus is nowhere more visible and illuminated than in border regions. Borders represent the international nature of the state; hence, they are outward looking. For example, many international borders today are a result of continually affirmed treaties between two or more nations.

Wilderness areas evoke the classic duality of nature and culture. As with frontier areas, the meaning of wilderness areas has undergone a transformation. Traditionally seen as zones of danger, sin, and desertion, today they are associated with the modern phenomenon of nature management and conservation. They have a unique feature: they are the only category in this class of areas that is marked, at least conceptually, by the absence of people (Oelschlaeger 1991; Neumann 2002). Perhaps because of this, they are associated with place-based transcendental meaning and spiritual experience. They represent not a struggle between civilization and barbarism, as frontier areas do, but rather refuge from civilization. Politically, however, they elicit a cultural clash between the ideologies of preservation and utilization. Many wilderness areas such as national parks and game reserves harbor important economic resources that are sought after by corporations, but many environmental groups led by ordinary citizens resist such a move. A prime example of this clash is the Arctic National Wildlife Refuge in Alaska, where many pro-utilization groups have been pushing the federal government to open the refuge for oil and gas drilling.

All of the above categories of areas are important to my discussion of Hunza. Hunza was once a frontier, a buffer zone, between the British and Russian empires. Today, it is marginal to the Pakistani nation-state, and it borders with China and Afghanistan. Further, many wilderness areas in the form of nature preserves have been established there. However, throughout its history Hunza has been predominantly described and defined by its geographical and physical *remoteness*, a word whose "semantic density" is associated with inaccessibility and lack of connectivity.[7]

Previous studies on Hunza and the surrounding region have focused not on remoteness but rather on one of the other categories. The region has been the focus of anthropological study by Western scholars since the 1980s. Studies focusing on the "local" have looked at the pre-Islamic history of the region, examining practices "left over" from that time (Sidky 1994; Parkes 2000; Csáji 2011). Representing broader shifts in anthropology, there is a growing literature on Hunza that examines the influence of outside political, economic, and cultural forces. There is work of uneven quality on colonial and postcolonial state formation and administration in Hunza by Stellrecht (2006) and Sidky (1996). Stellrecht's is a wonderful study of the political evolution of the Hunza state in the context of contemporary geopolitics. He argues that the Hunza state's transformation, from an unknown mountain hamlet in the mid-eighteenth century to a substantial regional power, can be understood in terms of its important position on the network of routes between South and Central Asia. Sidky has used the outdated Oriental despotism theory (Wittfogel 1957) to frame his study of the development of an irrigation network and the rise of the Hunza state. Other work focusing on the contemporary period has emphasized the intraregional cultural exchange of art, literature, and religious beliefs (Marsden 2008) and interaction between local people and global Ismaili institutions (Steinberg 2006). Economic and cultural geographers have shown that the region has been an important node on the north-to-west trans-Himalayan trade route that connected with the historical and legendary Silk Route (Kreutzmann 1991, 1993, 1995, 1998) and analyzed the political consequences in the region of geopolitical realignment of routes and roads

(Haines 2004). Their works have been very important in showing the region's historical connection with its neighbors and consequently illuminating the imperial and political formations surrounding Hunza. As important as this literature is, however, it overlooks the fact that, although the region is better connected with the outside world than it was during the colonial period, its remoteness, both imaginary and material, remains intact.

This literature on Hunza in the context of historical geopolitics and contemporary globalization represents an established trend in anthropology that seeks to dispel the notion that remote and marginal communities were as isolated and "stuck in time" as contemporary discourses have implied.[8] Structural Marxist analysis (Frank 1966; Wallerstein 1974) rendered the association of remoteness with isolation meaningless, showing how core and periphery were tied in an uneven structural relationship. If early twentieth-century anthropologists constructed remote places as entirely isolated and cut off, by the early twenty-first century anthropologists were trying to convince the world that remote places are in fact connected and always have been.

While I agree with this analytical point, I also argue that despite globalization and modernization, Hunza's remoteness persists, both physically and culturally. I am interested in understanding how, despite being engaged with the outside world, Hunza remains a "remote area." The concept of remoteness as I discuss it in this book is first and foremost geographical remoteness. All places and people must exist in some geographical space. Phenomenologically speaking, space lends form to our thoughts. Humans think with spatial forms: the signs and symbols of the material world are imagined and invested with meaning. The space of imagination constructed for such purposes includes appropriate signs that embody the spirit of those ideas. Difficult and long-standing questions are matched by images of hard-to-get-to lands inhabited by ancient peoples. For example, questions about human origins and the quest for an ideal society often take one into remote and faraway areas.

But the history and contemporary ethnography of Hunza that I will present suggest that while the geographical remoteness of Hunza is a given condition, the way it is imagined has depended upon who has done the imagining and within which political, intellectual, and cultural context. So, for example, to the British explorers of the nineteenth and early twentieth centuries, Hunza's inaccessibility meant mystery and danger, which spurred personal adventure, excitement for British officers, and expansion of the empire. In the 1950s, Western medical professionals and farmers visited Hunza, interested in the methods of agriculture that had survived there due to its remote location. They attributed the renowned longevity of the people of Hunza to its inaccessibility, especially its freedom from the influence of industrial food and its systems of production. In the 1970s, when Hunza became part of Pakistan, its physical remoteness and inaccessibility were depicted by the Pakistani state in terms of "primordial nationalism." In the 1980s, Western tourists described Hunza's geographical remoteness in terms similar to those used by the Westerners who had visited Hunza in the 1950s. Also in the 1980s, Hunza's geographical remoteness made it a perfect candidate for the creation of national parks. All of these different representations of Hunza, I argue, have been made within or in response to certain cultural projects of modernity, such as geographical exploration, creation of international boundaries and borders, nationalism, tourism, and environmentalism.

Remoteness and Affect

In his discussion of the production of knowledge about nature, Hugh Raffles (2002, 326) describes intimacy as an affect, as the "perpetual mediator of rationality." The word *remote* drives from the Latin *remotus*, the past participle of *removere*, to remove. It is, thus, a basic condition of positionality, or relationality, measured/evaluated in terms of distance—farness or nearness—both in real (geographical) *and* in conceptual space. The semantic density (of the experience) of remote areas thus implies, as mentioned earlier, inaccessibility and lack of connectivity, which are affects of estrangement. That is, one experiences estrangement when one encounters remote areas. Thus, remote areas are marked by a unique estrangement; remoteness is to space what strangers or unfamiliars are to social relations. But not all strangers are equal or equally strange. Writing about this, Zygmunt Bauman states:

> The "unfamiliars" come in a number of kinds, of unequal consequences. One pole of the range is occupied by those who reside in *practically* remote (that is, rarely visited) lands, and are thereby limited in their role to setting the limits of familiar territory (the *ubi leones*, written down as danger warnings on the outer boundaries of the Roman maps). Exchange with such unfamiliars is set aside from the daily routine and

normal web of interactions as a function of special category of people (say, commercial travelers, diplomats or ethnographers), or a special occasion for the rest. Both territorial and functional means of institutional separation easily protect—indeed, reinforce—the unfamiliarity of the unfamiliars, together with their daily irrelevance. (1990, 147–48)

Thus, according to Bauman, in remote areas the strangeness of the strangers and unfamiliars encountered there is of a special kind. I would add, following Ardener, that visitors are unsure, a little vague, about the real determinants of their strangeness, and this very condition makes the strangeness of remote areas of a different order. Ardener states that a remote area "produces that note of eccentricity and over-definition of individuality, if you like an over-determination—or to exaggerate slightly, a structure of strangers" (1989, 222). Da Col and Graeber further the concept of remote areas, building on Ardener, describing them as "singularities or pockets of social space inhabited by 'event richness,' conceptual vagueness or unusual boredom" (2011).

Building on this theoretical point, I argue that the history of Hunza shows that it has been represented not only with an affect of estrangement but also one that lies between the two poles of revulsion and enchantment.[9] As Ardener says, "There is clearly something in the idea that distance lends enhancement, if not enchantment, to the anthropological vision" (1989, 38). By "enchantment," I mean being smitten by an idea or condition and by "revulsion," I mean a negative view, most of the times irrationally driven, of an idea or condition. The people of Hunza have been depicted as savages and caravan raiders, as ecological noble savages, and as perfect specimens of human health. In between these extremes they have also been represented as irrelevant and inconsequential to wider politics and culture.

Socio-Spatial Domains

My attempt to create a social analysis through the concept of remoteness is based upon a number of analytical strategies. This involves introducing eight socio-spatial domains in which the coproduction of Hunza's remoteness is analyzed: exploration of space, categorization of space, boundary marking/delimiting territories, antimodernization and romanticism, nationalism, transhumance migration, international tourism, and environmentalism. Social theorists have argued that while space is produced in everyday human activity (Lefebvre 1991; Harvey 1999, 204), it is more than just a container for objects. Rather, it also subsumes relationships between objects that occur in a space. Building on this, I argue that a sociospatial domain is the relationship and hence meaning that is ascribed to objects in space and it is defined by the particular dominant discourse in which that space emerges. In all of these domains, Hunza geographical remoteness was coproduced with a particular discourse and attendant spatial practices. For example, the British imperial discourse was based on the idea of the spread of civilization, which produced the spatial practices of exploration and expansion in the nineteenth century. The discursive structure of British colonialism used Hunza's geographical remoteness as a justification for its civilizing mission. Clearly, I am not arguing here that remote areas caused the spread of civilization. Rather, they gave an additional dimension to the civilizing mission. Spreading civilization, among other things, meant greater contact and connectivity to those places that had been isolated. The British colonial explorers, travelers, and officers in India who wrote about their experiences for the British public coproduced imagined remote communities of uncivilized and civilizationless people as objective realities.

The link between civilization and remote areas is particularly illuminated in the genre of travel writings, in which the imperial civilizing discourse[10] is always the subtext. The demand for books written on the region surrounding Hunza, on the frontier, steadily rose through the nineteenth century in Victorian England. Mountstuart Elphinstone's *Kingdom of Caubul*, first published in 1815, was an instant hit in England, indicating an increasing appetite for knowledge and information about distant lands. As Huttenback writes about the people and literature of the Great Game era: "The curious breed of romantic 'men on the spot' who guarded the frontier marches did, indeed, believe themselves engaged in a struggle of cosmic significance, and in this illusion they were supported not only by the writers of fiction but by a press and public becoming ever more enamored of imperial grandeur and of the irresistible strength of Britain's arms and her national virtue" (1975, 2). By the end of the nineteenth century, various authors expressly acknowledged that such books were written, among other reasons, to introduce distant places to the general public back home. These places were often at the frontiers of the empires or interior hinterlands that had not been properly brought under control.

In 1900, Colonel Algernon Durand, the second political agent at Gilgit, the seat of British administration in the region, wrote in the introduction to his book *The Making of a Frontier*: "My reason for writing this book was that as the story of the development of Gilgit Frontier, told in my letters and diaries, was read with interest by some who saw those papers, it seemed probable that its publication might give to those who have no chance of seeing the sort of life their countrymen lead on an uncivilized frontier, a faithful idea of what such an existence means" (1900, x).

In 1896, in the preface to his book *The Heart of a Continent*, Francis Younghusband justified the necessity of a written account of his journeys in the northern frontier region of India: "To do this in conversation [give accounts of his journeys] is, in my case, a hopeless task—because, for one thing, my experiences of travel have now accumulated so heavily; and, for another, I find insuperable difficulties in giving by word of mouth accounts of travels in strange lands unfamiliar to the hearer. At the same time I am always experiencing the wish that my friends should be able to share with me, as much as it is possible to do so, the enjoyment I have felt in looking upon Nature in its aspects wild, in distant unfrequented parts of the earth, and in mixing with strange and little-known peoples" (v).

Colonel Thomas Gordon, a member of the Forsyth Mission of 1873 to the court of Yaqub Beg, the amir of Kashghar, traveled from Leh on the Indus across the Karakorum to the little-explored region of eastern Turkestan, engaged in map making for the mission and playing a vital role in the Great Game. He wrote: "My book . . . makes no pretension to be in any way a record of scientific exploration: it merely relates to what fell under 'every-day' observation. . . . The idea of writing it was suggested by my sketches forming such a complete series 'from the Indus to the Oxus' as to merit publication simply on the ground of representing to a very great extent life and scenery never before pictured. . . . The whole of the illustrations (with the exception of four colored plates) are facsimile copies of my sketches made on the spot" (1876, v–vi).

All of the above quotations illustrate that remoteness was conceived and constructed as a space worthy of being narrated and written about. Why did the British feel the need to tell stories of places that their compatriots "have no chance of seeing," "strange lands unfamiliar to the hearer," that had been "never before pictured"? On a personal level, writing such stories of explorations heaped prestige on the author. But on a wider social and political level, in telling stories of these remote places, Durand, Younghusband, and Gordon were inviting people back home to imagine these areas and their role in the spread of civilization, progress, and governance. Reporting from the remote corners of the empire, these authors gave the general public in England a chance to imagine an imperial vision in which remoteness stood as a sign of the vastness of the British Empire. The spatial practices of the explorers strengthened the discursive structure of the British Empire.

As the nature and intentions of British interventions and engagements evolved in Hunza from those of a neighboring power to an adversary power, evaluations of Hunza's remoteness and its people, both official and unofficial, also changed. The discursive structure was now based on the threat to empire and civilization from remote areas, a shift that also changed spatial practices. Remoteness was now produced as a space of danger, and spatial practices were carried out to subdue this danger through the slow invasion of the Hunza territory.

Likewise, in the mid-twentieth century, encounters between outsiders and the people of Hunza were framed within an antimodernist discourse in which Hunza's geographical remoteness became the location from which to launch a critique of modern industrial agriculture. In the 1970s, encounters between the Pakistani state and Hunza were framed in the domain of geopolitical nationalism in which the Pakistani state, using a primordialist discourse, linked Hunza's remote past with its modern form. Hunza was presented as a place "saturated with nationalistic meaning" (Kürti 2001), yet at the same time as a marginal place that needed to be brought into the mainstream economic and social development of the nation.[11] In the first decade of the twenty-first century, in the domain of environmental conservation, romantic discourses on Hunza's nature and culture frame encounters between the people of Hunza and outsiders and provide the justification for creating and setting aside (hence removing from human use) conservation areas such as national parks.

Now clearly, it is not the case that during a particular period Hunza was subjected to only one kind of spatial practice and the attendant discourse. Multiple practices and discourses existed side by side in Hunza. My methodology in choosing a particular discourse and spatial practice for a particular period is somewhat simplified, but not arbitrary. I chose these discourses and practices because they do, in an empirical sense, reflect representations of Hunza as a remote area at a particular time. Also, I chose discourses according to their importance in constructing *Hunza* as a remote place at a particular moment in Hunza's history. So, for example, the discourse of wilderness emerged elsewhere in the late nineteenth century but it did not become relevant to my story of Hunza until the late twentieth century.

Remoteness and Zomia

The historian van Schendel (2002) argued that the development of area studies after World War II meant that certain areas, organized around global political formations, became the focus of academic study. Areas marginal to those central areas of study were neglected; these were often mountainous or otherwise hard-to-reach regions populated by people who had, for various reasons, escaped the forces of centralized power. Van Schendel focused in particular on the Southeast Asian Massif, the mountainous zone between China and Vietnam, Laos, Burma, and Thailand, which he termed "Zomia," meaning the place of the highlanders. Scott (2010) develops this idea, arguing that the Southeast Asian Massif is an area where ethnic identities become fluid and new identities are formed, as the people living there develop strategies to persistently escape state power. Others have argued that the key points made by Scott do not all hold up under scrutiny—for example, Scott's notions of deliberate illiteracy, egalitarian political structures, and "escape" agriculture (Dove, Hjorleifur, and Aung-thwin 2011; Michaud 2010). As Shneiderman has argued in reference to her own fieldwork in central Nepal, "While for both empirical and political reasons the term Zomia itself may not be entirely appropriate to the Himalayan Massif, the analytical imperatives that underlie Scott's usage of it can be of great utility to those working in the Himalayan region, particularly the emphasis on the ethnic, national, and religious fluidity of highland communities, and their agency vis-à-vis the states with which they engage" (Shneiderman 2010, 290).

This conception of Zomia as a refuge from a central political power was suggested for the western Himalayan region by Graham Clark in 1977 when he asked, "If these Himalayan people are in some way acephalous, then the puzzles of boundary definition, and the corresponding problems of what it is that constitutes a people, would become more intelligible" (1977, 344–45). Clark was investigating the most recurring theme in Zomia studies: the invention and fluidity of ethnicities. Clark was searching for the origins of the "Dards," a group of people in the western Himalayas referred to by colonial writers as the original Aryan race. Although Hunza itself does not fit the description of Zomia because it had been a centralized polity since the seventeenth century, the region surrounding it does. But more important than the formal similarities are the substantive similarities in Scott's observation on Zomia and what I found in my research on Hunza. These are the strategies that the people of these places employ in their engagement with external powers. My study explores not only how the idea of remoteness is constructed through the material practices and discourses of powerful outsiders but also the ways in which the people of Hunza creatively engage with the notion of remoteness that is applied to them, using outsiders' perceptions to further their own interests.

Studying Hunza

Hunza today is a district of what is now called the Gilgit-Baltistan region of Pakistan. The district boundaries are the same as those of the former Hunza state. The total population of Hunza district is about seventy thousand people, divided among three main valleys and into three ethnic groups.[12] The total area of Hunza is roughly five thousand square miles. The landscape of terraced agricultural fields interspersed with fruit orchards testifies to the still predominantly agrarian base of the local society. More and more people, however, are now engaged in off-farm work in government, private, and nonprofit sectors.

The lower Hunza valley (in the south) is inhabited mainly by the Shinaki ethnic group, whose members speak the Shina language. The central Hunza valley, the former seat of the rulers of Hunza and the region popularly referred to as "Hunza," is inhabited by the majority: Buroshiski-speaking Burosho people. Buroshiski is considered a language isolate with no connection to the neighboring Indo-European or Indo-Iranian languages. The Burosho people are thought to have come from Iran in the twelfth century. During the days of the Hunza state, they formed the elite and noble classes of Hunza, and the more powerful families were generally exempt from state taxes. Buroshos today are heavily represented in the several development and conservation NGOs working in the region, and many are also employed as soldiers in the Pakistani Army. Karimabad, the former capital and now the hub of economic activity in the region, is located in central Hunza and has become a popular tourist destination for both local and international tourists, with an ever-growing number of hotels, cafés, and shops.

The northern part of Hunza valley, Gojal, is inhabited by the Wakhi people, who speak Wakhi, a dialect of Persian; they number a total of about ten thousand in all of Pakistan. Gojal was historically famous for its lush and expansive pastures, ideally suited for keeping large herds of livestock and harboring good game, but also known for its harsh weather. Historically, economically, and socially marginal, the upland Wakhis are considered a simple people by Buroshos of central Hunza. But in reality, Wakhis today are highly educated relative to the rest of the country, having taken advantage of education opportunities made available through development projects, and are engaged in various off-farm activities such as trade and tourism. Some of the most famous mountaineers of Pakistan come from this region, particularly from the Shimshal valley, the site of my most detailed ethnographic fieldwork. In the days of the Hunza state, the Shinaki and the Wakhi people were taxed disproportionately higher than the Buroshos of central Hunza.

Except for a handful of Shia villages in the central Hunza valley, all the inhabitants of the state of Hunza are Ismailis.[13] Ismailism is an offshoot of Shia Islam. Its adherents follow a living and present imam as their religious and political leader, whom they believe to be the only valid interpreter of the true meaning of Islam. The present imam, Prince Karim Aga Khan, lives in Aiglemont outside Paris, France. In addition to his role as spiritual leader of the Ismailis, he is head of a number of philanthropic institutions, runs a large international educational network, and is perhaps best known in the West as a wealthy, Harvard-educated owner of racehorses.

The historical aspect of the book focuses on the central Hunza valley and the seat of the Hunza state, while the contemporary ethnographic research centers on Shimshal, a village in the northeastern corner of the state populated by Wakhi speakers. Shimshal is remote to Hunza, and Hunza was remote first to the empire and later to the Pakistani state, providing a common theme of remoteness through the work. Obviously, besides being populated by different ethnic groups, the two are different in other ways: Hunza is a bounded political entity, enjoying sovereign rule over a large territory and population, compared to Shimshal, a village that has never had independent political status. Shimshal's status as a margin within a margin, and indeed a remote area within a remote area, adds to the complexity of the story and the analysis.

Despite the differences in the nature of the two as political entities, their historical representations share striking similarities. During the period of Hunza's history I study (the late nineteenth century until the early 1960s) and that of Shimshal (the 1970s and 2003), neither place was accessible by motorized transport; the only way in or out was an arduous journey by foot or on a horse or other animal. Each place has attracted considerable attention from the outside world, despite or most likely because of its remoteness. Both have been considered Shangri-las, mystical, paradisiacal valleys, and both have had similar relationships with external power centers. John Mock (1998) and David Butz (1992, 1996, 2000) have done significant research in Shimshal. The former studies indigenous conceptions of nature and spirituality, placing those conceptions within the broader community of the Pamir region, and the latter investigates conceptions of space and identity among the Shimshalis. My work has benefited from their scholarship.

Anyone who works with empires' historical accounts of societies they have ruled comes up against the problem of relying too much—or sometimes only—on texts, archives, and memoirs produced by representatives of the empires. The only local history produced on Hunza was written under the guidance of a former British official in the region (Beg 1935). Separating bias from facts is a challenging exercise. Invariably during the course of my historical research, I sometimes found myself dealing with only one text on an issue or event. Bernard Cohn has argued that texts must not be seen as facts but rather be understood in terms of the meanings intended. He states, "This can only be done through understanding the shadings of language and the structure of the text, and through the development of sensitivity to changes in form through time" (2004, 48). The choice of words in colonial texts about Hunza shows such a relationship between language and meaning. For example, the British described Hunza's raids on the Kirghiz herders as "caravan raiding." The word *caravan* here was (perhaps deliberately) used to conjure up in the reader's mind the trade caravans that traveled between India and Central Asia. In fact, the raids were made on Kirghiz camps to collect taxes owed for their use of Hunza's grazing lands.[14]

It is, of course, impossible to relive the experience of visitors to Hunza and those they visited at this time. As Hastrup states, "We will never know the unequal experiential densities of categories and social spaces through hearing and seeing, because the social is no mere text but a lived reality" (1989, 227). The closest I could come was to relive, to some extent, the social world of remoteness through ethnographic fieldwork. By choosing Shimshal as the site, I was trying to recapture the feelings that my historical subjects may have had, and in so doing also following the traditional anthropological craft—that is, going to the most remote and isolated place possible. An added twist is that I am, as a Pakistani, a "native" anthropologist. Although this meant that there was no language barrier between me and the educated people with whom I spent time and we shared experiences of religion and nationality, Shimshal was in many ways as distant from my life experience as anywhere in the world might have been.

Chapter Outline

The construction of Hunza's remoteness in each of the socio-spatial domains introduced earlier is the topic of the eight chapters of this book. Each chapter deals with a particular domain and the accompanying discourse and spatial practices in which Hunza's remoteness is constructed and reinforced.

In the first chapter, I look at the construction of Hunza's remoteness in the socio-spatial domain of geographical exploration. I show that the explorations and surveys of the region resulted not only in acquisition of geopolitical information, including descriptions and cataloguing of people and space but also in an exaggerated sense of uniqueness; in the British account, the region was the origin of sacred rivers and of people who were the ancestors of both Indians and Europeans. That is, the people of this region were viewed by the British as the original specimens of the Aryan race. In their explorations of the northern frontier region of the Indian subcontinent, the British deployed the discourse of "lifting the veil" to construct the region as full of mystery and mystique and, hence, an appropriate and suitable candidate for exploration.

In chapter 2, I look at the construction of Hunza's remoteness in the domain of categorization and governance of imperial territory. I look at how the use of discourses of "friction of distance" and "rhetoric of distance" in this domain renders Hunza a remote place, inhabited by primitive and savage people. The resistance of the mir of Hunza to British frontier policy and his political machinations with the Russians and the Chinese made him a quintessential savage who purportedly did not understand the benefits of civilization, which the British thought they were bringing to Hunza. The people of Hunza were constructed as caravan raiders who wreaked havoc on "trade" in the remote frontier region. The backwardness of Hunza, however, was not of a regular kind. Rather, these people were not just inferior but also beyond reason; they were the radical other. While the British viewed the people of Hunza as remote and isolated, the rulers of Hunza positioned themselves differently: at the center of three powerful empires rather than at the edges of civilization.

I then look at Hunza in the postconquest era of 1891, when it became the northernmost outpost of the British Empire. Using frontier settlement and administration as a socio-spatial domain, I look at the ways in which irrelevance and lack of urgency in the frontier settlement process constructed

Hunza's remoteness. Yet the society of Hunza was also treated by the British colonialists with an air of exclusivity and a tinge of romanticism such as distant and quaint places often are associated with. After the conquest, Hunza became an ideal society in need of British protection and benign paternalism. Hunza's remoteness was constructed in this discourse in requests by British officers for reenactments of caravan raiding and rerepresentations of the travel experience to Hunza from Kashmir. The mir of Hunza, now fully loyal to the British, consolidated his power under British rule and extended his authority over areas where he previously had none.

In the fourth chapter, I explore the construction of Hunza's remoteness within the discourse of antimodernism. In the mid-twentieth century, a number of Western medical doctors and farmers visited the region, fascinated with the remarkably good health of the people of Hunza, which they attributed to the traditional methods of agriculture and food production that had survived because of Hunza's isolation from the modern world. If Hunza's location was earlier seen as being on the margins of—and even beyond—civilization, now it was seen as a *refuge from* civilization. But it was no ordinary refuge. It was depicted as a rural utopia, a Shangri-la. As in the past, the mir of the time manipulated outsiders' representation of his domain as remote by playing along in a hopeless effort to strengthen his dwindling position against the Pakistani state, which had its own policies that reinforced Hunza as remote, even as it tried to connect Hunza more closely with the mainstream society and economy.

The fifth chapter deals with Hunza's postpartition history and its representation as a marginal but remote place within Pakistani geopolitical nationalism. I look at how discourses of primordialist nationalism and development constructed Hunza. Hunza was represented as a region that harbored the origins of the mythical Pakistani nation. It became the indigenous face of the modern Pakistani nation-state. The chapter also looks at how the development discourse of an international NGO divides the territory of the erstwhile state of Hunza into central and remote areas and how this division is perceived by its inhabitants.

In chapter 6, I examine how indigenous notions of space and place are structured and how the people of Shimshal act and perceive their geographical remoteness. The Shimshali seasonal migration of yaks creates zones of remoteness within their own cosmologies. Here remote space is constructed through internal mobility that is a function of the location of grazing areas and migration routes and the behavior of the yaks. I look at how Shimshalis construct an indigenous sense of remoteness through discourses of separation and integration in the socio-spatial domain of transhumance migration and pastoralism, which forms the basis of their subsistence and hunting practices. I show that Shimshalis feel that despite the lack of connectivity with the outside world, they have become increasingly vulnerable to the flow of ideas and material from the outside.

In the seventh chapter, I explore the encounter between tourists and the people of Shimshal during the last two decades of the twentieth century, within the domain of global tourism and mediated by a discourse of cultural authenticity and hospitality. This discourse is similar to the one discussed in chapter 4, when Hunza was represented as Shangri-la. At that time Hunza's remoteness was framed as an explanation for its preservation of pristine agricultural practices; by the early 2000s, Shimshal's remoteness was seen as an explanation for the preservation of its culture of hospitality. One aspect of tourists' experience was that the Shimshalis "performed" the ideal host, welcoming tourists into their homes. Tourists believed that increasing accessibility to Shimshal by road threatened the village's remoteness and thus its very culture.

Considering the socio-spatial domain of environmental conservation, in the final chapter, I look at the construction of Shimshal's remoteness in two apparently disparate romantic discourses on nature. In one discourse, Shimshal appears as the last refuge of endangered species such as the Marco Polo sheep and the snow leopard. Shimshal's vast pastures are rendered empty and pristine in this discourse, evoking the conventional image of the area as a zone far from the modern forces of industrialization, capitalism, and state expansion where nature still reigns. The people of Shimshal appear in this space as out of place. In another discourse on nature, however, Shimshalis appear as an integral part of it and, like nature itself, they also are uncorrupted and untainted by the ethos of modernity. I show how the Shimshalis try to articulate their identity with this latter discourse in the context of the establishment of Khunjerab National Park in 1975, which set in motion a long and still simmering conflict between them and the Gilgit-Baltistan Forest Department.

In the epilogue, I draw together the various aspects of remoteness that I have discussed in the book. I look at how the existence and perceptions of remoteness and remote areas have become inherent conditions of modernity and the process of modernization. I look at how some of the historical meanings associated with remote areas resurface in the contemporary "war on terror" in the region. I conclude by arguing that despite globalization and the evolution of far-reaching technologies, both in a literal and metaphorical sense, the meaning of remote areas is changing yet again. Today, remote areas are described less by their accessibility and familiarity and more by their anonymity.

CHAPTER 1

Lifting the Veil

The Sacred and Political Geography of Hunza

Throughout the second half of the nineteenth century, the British and the Russians were locked in geopolitical rivalry that came to be known as the Great Game. The term refers to the struggle between the two powers to mark their spheres of influence in the countries along the northern borders of the British Indian Empire. The British became alarmed by the slow but steady expansion of tsarist Russia eastward into Central Asian khanates to the north of Afghanistan and into the western edge of the Chinese Empire. Although the British never objected to Russian conquest of these Central Asian khanates, which paid allegiance to the Ottoman sultan, they did worry about Afghanistan, farther to the south and east: they did not want Afghanistan to fall to the Russians. The British had left Afghanistan in 1842 and they feared that a Russian foothold in Afghanistan would mean a threat of Russian invasion of British India, the crown jewel of Britain's imperial possessions.

Earlier in the nineteenth century, in response to the perceived or real threat of a combined Franco-Persian attack, the British had already expended a significant amount of political and diplomatic energy in extending their sphere of influence to local societies and certain strategic mountain passes, such as the Khyber Pass near Peshawar and the Gomal Pass near Quetta, on the northwestern frontier. The British now looked toward the region north of Kashmir beyond the great Himalayas and the mighty Indus, up to the Central Asian watershed, as a possible route by which the Russians could invade India.[1] The region is the meeting place of the world's three

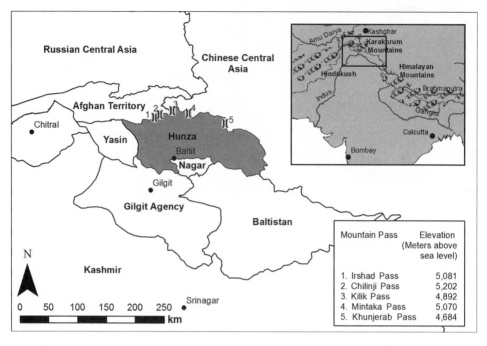

Figure 1: Map of Hunza and its northern frontier, circa 1880 (map drawn by Chris Milan)

highest mountain ranges—the Hindukush, the Karakorums, and the Pamirs. The maze of remote and inaccessible mountain valleys became a field for the political machinations and social imagination of the Great Game between the British, Russian, and Chinese empires as well as various smaller players who were drawn into the struggle (Alder 1963).[2]

By the mid-nineteenth century, a primary geopolitical goal for the British became defining and delimiting the territorial boundary of their empire in this trans-Himalayan region through mapping and explorations in order to thwart Russian designs on the area. The Russian expansion in Central Asia ushered in an era of British exploration of the Hindukush, Pamir, and Karakorum watershed that divided South and Central Asia. It was around this time that the British public and colonial officers, especially those belonging to the East India Company, were beginning to grasp Hunza in their imagination as distant yet geopolitically important. In 1926, George Curzon, the viceroy of India from 1899 to 1905 and architect of India's northwestern frontier policy, wrote this about the state of Hunza: "Eighty years ago, when Golab Singh, the Raja of Jammu, to whom the British Government had just given, or rather sold, Kashmir, was endeavoring to subjugate and define the outlying and Trans-Indus portions of his new possession, and when neither he nor the Indian Government knew very clearly where that border was, we catch a glimpse of Hunza-Nagar states" (183).

The British knowledge of the physical geography of the region, however, was limited. On its northern border there are about half a dozen mountain passes through which an invader could enter.[3] As Curzon stated, "This remote mountain valley has an importance for Englishmen which its geographical isolation would lead few to suspect. It is one of the northern gates of India, through which a would-be invader must advance if he advances at all" (1926, 149). The British were latecomers to the region, and the northern frontier of Kashmir remained relatively uncharted territory till the 1860s, when the Great Trignometrical Survey was done (Edney 1997).[4] They did, however, have some vague idea about the sacred geography of the area through their readings of either primary sources such as the ancient Brahmanical texts or derivative work of scholars of the East India Company. In the late eighteenth and early nineteenth centuries, Orientalist[5] scholars of the East Indian Company (EIC), such as William Jones, Nathaniel Halhed, and Henry Colebrook, had studied and translated sacred Sanskrit texts of the Brahmans for the company's administrative purposes (Cohn 2004; Inden 2000).[6] Jones and other Orientalist scholars of the EIC became interested in references to a golden civilization that had once existed in the ancient past somewhere in the region north of India and west of China (Bryant 2001; App 2009). The stories referred to a sacred river that flowed from the base of a mountain around which this civilization flourished. These scholars postulated that this was the civilization of Indo-Europeans, the common ancestors of Europeans and Indians.[7] Some argued that these Indo-European people were the original Aryans, the predecessors of those Aryans who came to India in the post-Harrapa period. Similar stories and tribal names also appeared in the classical Greek history familiar to the Orientalist scholars. They looked for a common structure and meaning in Latin, Greek, Persian, and Sanskrit texts and languages as proof of the existence of a proto-language spoken by these people. Jones even considered Sanskrit to be the language most closely related to this original one (App 2009, 42).

This Orientalist narrative caught the imagination of members of the Royal Geographical Society and others in the geographical explorers' community who led efforts to map and survey this region. They determined that the common ancestors of Indian and European people had lived in the

Pamir region on the northern frontier of Hunza state, in what is known today as the Wakhan corridor in northeastern Afghanistan.[8] Their exploration focused on two things: charting the physical geography of the region, especially the condition of mountain passes, and searching for the source of the river Oxus—the location of an ancient Aryan paradise.

This intersection of British interest in the region's geopolitics and its mythology was not contingent. Rather, the British fostered this somewhat out-of-fashion Orientalist narrative to energize their explorations aimed at creating a boundary of their empire. As David Harvey states, "Geopolitical conflicts often imply a certain aestheticization of politics in which appeal to the mythology of place and person has a strong role to play" (1999, 209).[9] The use of the appellation "Great Game" to represent a geopolitical conflict indicates that the appeal to players was not merely political but also involved interest in the myths of the region in which it was played—in this case its extreme remote location.

In the second half of the nineteenth century, the predominant Orientalist influence on British exploration accounts of the region began to be replaced by Darwin's theory of evolution. This shift was not very sudden; for a time the Orientalist myth of an original Aryan race existed side by side with Darwin's theory of geographical race. But the gradual decline of the influence of Orientalism on British policy accompanied the rise of a more imperial and masculine form of British colonialism.

Most important for our purpose here, the credibility of both of these narratives depended on remoteness. The geographical isolation of the region provided the credible spatial setting in which both the Orientalist narrative of an original but now forgotten and hidden Aryan race and the Darwinian narrative of a biological evolution based on geographical isolation made sense. In addition to this spatial setting, remoteness also bestowed prestige and status on the explorers who succeeded in reaching the region.

There are two important metaphors that aptly describe the way the British framed their explorations and journeys in the region during the period between the 1830s and 1890s: "lifting the veil" and "blank on the map."[10] Taken individually, these metaphors represent the form of colonial political rhetoric and the nature of British knowledge of the region respectively. Read together, they represent the process of imperial territorial expansion in faraway places and its justification as the spread of civilization as well as the role played by geographical knowledge in that process.[11]

These metaphors were related to the British exploration in another way. Their power in exploration narratives is due to the values they ascribe to the domain and practice of exploration: challenge and discovery. These values are held together in imagination by a common thread of remoteness. And it is remoteness that gives these two central values of geographical exploration their meaning. Relatively more truth value and symbolic status is attached to the exploration of geographically remote and inaccessible areas because remoteness signifies distance and hence invokes other categories associated with challenge and discovery, such as endurance, risk, estrangement, and fascination.[12] The remoteness of Hunza meant that knowledge of it was beyond the everyday knowledge of the colonial state, but it was nonetheless within the horizon of exploration. The colonial state supported exploration, private and official, while explorers provided knowledge and lifted the veil from this remote region.

"Blank on the Map" and "Lifting the Veil"

Although the wording of the metaphors "blank on the map" and "lifting the veil" did not appear as such in the colonial geographical and speculative accounts of the region, various other expressions meaning essentially the same thing appear in the British imperial language of frontier expansion and military conquests of the region in the mid-nineteenth century. The knowledge of Himalayan geography produced because of the Great Game is well captured by Trench's metaphor: "During the last 20 or 30 years . . . , the rays of European light and enquiry have, after so long an interval of night, *begun to light up the darkness of Central Asia*; and owing chiefly to the gradual approach of the Russian frontier from the North to the English boundary on the Himalayas, immense additions have, by the exploration and surveys of British and Russian officers, been made to our general geography of Central Asia" (1869, 34; emphasis added).

A decade later, expounding on the role of geopolitical conflicts in the production of knowledge of the "least accessible" regions, John Biddulph, the first British political agent to Gilgit, wrote, "Within the last half a century, war and private adventure have contributed so largely to *making known the least accessible regions of the continent of Asia*, that few parts remain of which a fair general knowledge does not exist" (1880, 1; emphasis added).

These expressions of "lighting up the darkness" and "making known the least accessible regions" were used to denote the accumulation of both the geographical and the historical/ethnological knowledge of the region. The forays made by the British in the region did not seek only to explore the mountain passes at the Karakorum watershed; they were also conducted to verify the significance of the region as the center of the universe,[13] as mentioned in ancient classical texts (Yule 1872, xxi–xviii). Accordingly, almost all the explorers, officials, and scholars interested in the region speculated that the region north of Hunza where the Oxus flows is the primeval paradise mentioned in both Eastern and Western classical texts. A British Army officer stationed on the Afghan frontier during the mid-nineteenth century wrote in 1869 on the question of this region: "Probably the principle of *omni ignotum pro magnifico* has never been more fully exemplified than in the interest which this region of the world has, for a great number of English geographers and ethnologists, long possessed. As regards the former, the *mystery which so long enveloped the whole subject* naturally whetted their desire for the acquisition of more accurate knowledge, while to the latter, the origin and history of the primeval races who had here their dwelling, and the successive changes through which in the course of bygone ages they have passed, has always presented a very wide field of enquiry" (Trench 1869, 33; emphasis added).

Scottish Orientalist scholar Henry Yule took this idea further in his preface to a new edition of John Wood's *A Journey to the Source of the River Oxus* (1840), the first comprehensive colonial account written on the Oxus. Yule, himself an explorer, the president of the Royal Asiatic Society, and an officer in the British Indian Army, wrote about the importance of the region as the origin of human civilization:

> Few regions can present claims to interest and just curiosity so strong and various as that heart of Asia which gives birth to the Oxus. Forming the great physical and political watershed of the old Continent, *it has been through difficulty of access long shrouded in mist that has darkened and distorted a variety of ethnological and geographical problems, mist which only now begins to lift; it is a center of primeval tradition* as well as of modern theory regarding the primitive history of mankind. . . . Here is the one locality of earth's surface to which, if some interpretations be just, the Mosaic narrative points, in unison with the traditions of Aryan nations, as the cradle of our common race. (1872, xxi; emphasis added)

If for Trench, writing in the 1860s, the region was still enveloped by mystery, for Henry Yule, writing in the 1870s, the mist under which the region was shrouded was beginning to lift.[14]

Yule's reference to the "Mosaic narrative" is the biblical version of history, whereby the descendants of Noah, having built the tower of Babel, were scattered throughout the earth speaking different languages. Yule and others like him seemed to see the origin of the Oxus as connected with this original people. We cannot be sure that Yule himself believed the biblical narrative, given the prevailing challenges to that narrative from Darwin's evolutionary theory, but this practice of reading biblical story into the history of the people of the world was a familiar Orientalist strategy. Yule's contemporary Henry Rawlinson, a giant of that special breed of explorer and Orientalist and president of the Royal Geographical Society from 1874 to 1875, felt that the people surrounding the Hunza region had significance in the sacred texts of the Hindus themselves. Rawlinson thus established some correlation between the biblical and the Brahmanical narratives.

> The present inquiry naturally opens with a glance at the antiquities of the region in which the Oxus rises. There is no need of an elaborate discussion, for the subject has been extensively, not to say exhaustively treated by Humboldt and Ritter; but I shall desire to draw attention to certain points which do not seem as yet to have received sufficient consideration. The region, then, I may say, which embraces the headwaters of the Oxus is not of less interest geographically and politically than it is on the account of its connection with the primitive traditions of the Aryan race. Whether Bounouf may or may not be right in regarding the term Pamir as a contraction of Upa Meru ("the country above Mount Meru") and is thus associating the name directly with the holiest spot in the Brahmanical cosmology, it is certain that the geographical indications of the Puranas do all point to this quarter of Central Asia as the scene of the primeval Aryan paradise. (1872, 489)

Even George Nathaniel Curzon, a prolific scholar and viceroy at the turn of the twentieth century who, as we shall see later on, was heavily influenced by the Darwinian narrative, speculated on the region using Oriental discourse. For example, to Curzon the Oxus was "that great parent stream of humanity, which has equally impressed the imagination of Greek and Arab, of Chinese and Tartar, and which, from a period over three thousand years ago, has successively figured in the literature of the Sanskrit Puranas, the Alexandrian historians, and the Arab geographers. . . . Descending from the hidden 'Roof of the World,' its waters tell of *forgotten peoples*, and whisper secrets of *unknown lands*. They are believed to have rocked the cradle of our race" (1896, 15–16; emphasis added).

All three scholars quoted above reproduced the claims made by the famous Orientalist scholar William Jones of the East India Company a century before. The construction of the whole region as a remote place was tied to the pursuit not merely of difficult geographical knowledge but also of mysterious ethnological knowledge. The geographical search for the source of the river Oxus and an Aryan race by the British explorers in the late nineteenth century was the mirror image of the research on ancient Hindu texts by the scholars of the East India Company a century before. In both cases, distance lent authenticity to their crafts and claims: the authenticity of a text was judged by its age (distance in time), while the authenticity of claims of sources (of rivers) was judged by remoteness (distance in space).

Exploration of the Oxus

With the colonial expansion in the nineteenth century, geographical societies were established in a number of major European cities. Their purpose was to provide more accurate scientific data for the better management of newly acquired foreign dominions. In the late afternoon on May 24, 1830, several gentlemen gathered at the Raleigh Traveler's Club of London to launch the establishment of a society for the promotion of geography, which would become known as the Royal Geographical Society (RGS). The society's first president declared that its "*sole* object should be promotion and diffusion of that most important and entertaining branch of knowledge, Geography. . . . The interest excited by this department of the science is universally felt; . . . its advantages are of the first importance to mankind in general, and paramount to the welfare of a maritime nation like Great Britain, with its numerous and extensive foreign possessions" (Barrow 1831, vii; emphasis added).

In the second half of the nineteenth century, the RGS became one of the most important advocates of exploration of India's northern frontier and indeed the rest of the British Empire. Robert Stafford (1989), in his study of the rise of the geographical societies and geological sciences and the expansion of empire, has shown how the members of societies like the RGS played an important role in maintaining imperialism while also enriching themselves. He states that many of the RGS's geographers made clear the relationship of mutual service they hoped would prevail between their organization and their country's imperial establishment.

Between the 1840s and the 1920s, the RGS published countless scholarly papers on the geography and ethnology of the upper Oxus region. Contributors included Great Game players such as Francis Younghusband, Algernon Durand, and George Curzon, who were instrumental in settling this frontier region and its boundaries.[15] One of the first journeys to the river Oxus had been made in 1808, before the establishment of the RGS, by one Lieutenant John Macartney, a cartographer who served at the Peshawar embassy under Mountstuart Elphinstone in 1808–9 and who was the envoy of the East India Company to Afghanistan. In 1815, Elphinstone published his famous work, *An Account of the Kingdom of Caubul*. In the appendix, he includes an extract from Macartney's memoirs relating the lieutenant's account of his travels to the region in order to map the routes.[16] About the source of the river Oxus, he writes:

> It issues from a narrow valley two or three hundred yards broad in Wukhan, the southern boundary of Pamer. This valley is enclosed on three sides by the high, snowy mountain called Pooshtikhur, to the south, east, and west. The stream is seen coming from under the ice, which is stated to be at least forty spears in depth. The spring itself could not be seen in consequence of the great mass of ice formed over it, but there can be no doubt of the spring's being on this hill under the ice, for it does not appear that there was any open or break in any of the three sides mentioned by which it could come from a more distant point. I, therefore, conclude that this is the true head of the Oxus. (408)

Macartney identified a glacier at Lake Chakmak in the Pamir Mountains, in the Little Pamir,[17] to be precise, as the true source of the Oxus, a claim that would not be proven until some eighty years later by Lord Curzon.

The second British explorer sent by the colonial administration specifically to look for the source of the river Oxus was John Wood, then a captain in the Indian Navy, who undertook the journey in 1835. Wood traveled on a streamer up the Indus to Kohat, then took a boat up the Kabul River, then traveled overland into Kabul, Afghanistan, and entered the Badakhshan region, which was a semi-autonomous state at this period. As Wood was about to enter the Wakhan region of Pamir, a high tableland from which radiate some of the highest mountain ranges in the region, including the

Karakorums and the Hindkush to the south and Kun Lun to the north and west, he finally had a glimpse of the Oxus. Traveling further northeastward through the town of Kila Panj, which stood on the right bank of the Oxus, Wood noted that the stream split into two, with one branch going east and the other north. "The former [coming from east], we were told, conducted into Chitral, Gilgit and Kashmir, and the latter across the table-land of Pamir and Yarkand. I had now to ascertain, if possible, which of the two streams I was to trace. One of them it was certain must lead to the source of the Oxus, but which of the two was a question of difficulty" (1841, 217).

After Wood measured the relative temperature, depth, and velocity of the two streams, his first instinct was to follow the stream coming from the east. However, he took the advice of local informants seriously and eventually decided to follow the northerly stream. It led him to Lake Sarikol, which today runs along the border between Afghanistan and Tajikistan, some ten miles north of Lake Chakmak, which Macartney had named as the source (in Elphinstone 1815). Wood concluded that the source of the Oxus was not Lake Chakmak but rather Lake Sarikol on the Big Pamir. He named the lake Lake Victoria after Queen Victoria, and in 1841 the RGS awarded him its Patron's Medal.

The search for the source of the river Oxus gained momentum as the Russian frontier moved further east in the Pamir region in the 1860s. In 1868, the RGS funded a trip by George Hayward to the Yarkand and Pamir region to collect geographical information about the mountain passes for the purpose of defense (1870, 41). Hayward began by traveling to the foot of the northern passes of Borighil and Mintaka on Hunza's frontier. From there, he was to travel to the source of the Oxus, but he was murdered at Yasin, just miles short of his destination. (The events surrounding his murder are murky, but the British suspected the ruler of Yasin had ordered his murder at the behest of the Kashmir rulers, who actively discouraged British forays into the region.) At about the same time, the Survey of India sent a handful of "native" spies,[18] known as Munshis or Pundits, to trace the source of the river Oxus, and their accounts were published in the RGS journals (Montgomerie 1871, 1872). In 1872, members of the RGS such as Yule and Rawlinson, quoted earlier, ratified Wood's claim that Lake Victoria on Big Pamir was the source of the river Oxus.

British exploration of the source of the Oxus and geopolitical imperatives of the empire in the region continued to reinforce each other throughout the 1870s. In an effort to increase its political influence in Chinese Central Asia, in

1873 the government of India sent a second mission led by Douglas Forsyth to establish diplomatic ties with the newly independent state of Kashgaria in Chinese Turkestan (Forsyth 1875).[19] As part of this mission, one Captain Trotter traveled beyond Yarkand to the Pamir region to confirm the location of the source of the river Oxus. Trotter concluded that the source was Lake Chakmak in Little Pamir, identified by Macartney many years earlier, not Lake Victoria in Big Pamir, as Wood had claimed and Yule had ratified. Trotter argued that the original stream of the Oxus was the one that left the lake from its eastern shore. He was also aware of what that meant geopolitically, especially when demarcating boundaries and spheres of influence between the British and the Russian empires in the region. He wrote: "The importance of this discovery in the then state of political geography was very considerable, on account of the disputed boundary between Russia and Afghanistan, in the neighborhood of the upper waters of the Oxus" (1899, 442).

Still, the matter was not settled, and the search for the source of the Oxus continued for the next two decades. And so did negotiations between the Russian and the British empires about the boundary demarcation. Between the 1860s and the 1890s, a handful of other explorers (Veniukof and Mitchell 1866; Dunmore 1893; Roberts et al. 1896) speculated on the source of the river Oxus. But the matter got most attention in 1894, when Curzon, the future viceroy of India, traveled up through Hunza and Nagar territories to the Little Pamir and reached Lake Chakmak. Curzon claimed that the source of the Oxus was not the stream that left the eastern end of the lake, as Trotter had suggested, but rather a glacial cave formed by the Wakhjir glacier. A stream issued from the cave, flowing in the westerly direction without ever entering the lake (1896, 54).

One can easily read into this claim Curzon's inclination to pinpoint a definite starting point independent of a lake. The problem with a lake as a source is that many streams flow into a lake but more than one stream could be flowing out, thus making it difficult to claim a definitive single origin. Curzon's claim seems to have proven the most acceptable, as soon afterward the boundary was demarcated at this point, making moot the issue of the true origin. All in all, Macartney in 1808, Trotter in 1874, and Curzon in 1894 identified the source of the Oxus on Little Pamir, about ten miles, as the crow flies, north of the Hunza-Afghan frontier in Afghanistan.

Wood, Hayward, and Curzon received medals from the RGS for their explorations of the source of the Oxus. They asserted their claims—whether they identified the true source of the Oxus as Lake Sarikol, Lake

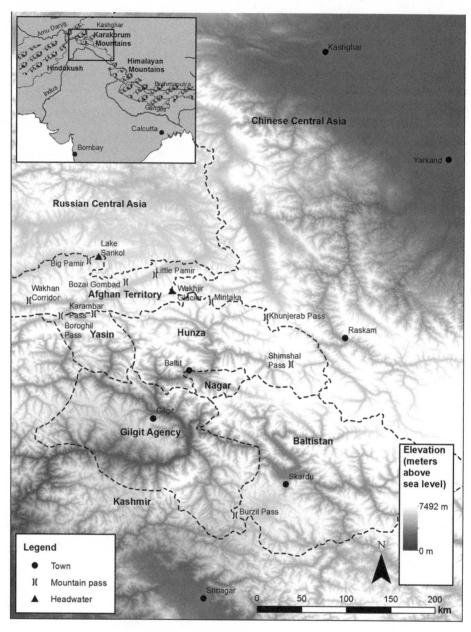

Figure 2: Map of Pamir Mountains and the river Oxus (map drawn by Chris Milan)

Chakmak, or a cave at Wakhjir glacier—in the pages of newspapers, books, and journals or in lectures they gave at venues removed from the physical domain of their explorations.

The Pamir Boundary Commission was the body that eventually demarcated the boundary between Russia and Great Britain in 1895.[20] The two lakes on Big and Little Pamir—Victoria and Chakmak—became the northern and southern boundaries of Afghanistan dividing British India and Russia (Todd 1951, 76–77). The northern passes of Hunza were to serve as the southern boundary of Afghanistan, with the British Empire thus creating a buffer zone of the Wakhan corridor between the British and the Russian empires. Geopolitically, a boundary at the source of the Oxus pushed the British sphere of influence on the north of the Central Asian watershed, a position that was sought after by the British. It is important to note that the commission placed no special emphasis on which among the various claims was the true source. The author of the report for the British wrote this about the source of the Oxus:

> The Agreement of 1872, concluded at the period when Chinese Power had temporarily disappeared from Yarkand and Kashgar, and the discussion of exact limits in these regions seemed outside the range of practical politics, proved twenty years later difficult in application to the actual geography, where, three, or perhaps four, sources might each claim to be regarded as the fountain head of the mighty Oxus. . . . Judging from the respective sizes of the Wakhan Daria at Sarhad, 200 miles, and of the Aksu/Murghab at Pamiriski Poste, 150 miles from their point of junction, I should be disposed to say that the former, whose source is at Wakhjir, below Wakhjir Pass, is the more important stream. So choice of the intermediate Victoria branch was the best possible compromise under the circumstances. (Gerard 1897, 1)[21]

In the end, then, it was just a "best possible compromise" that established the source of the Oxus. The actual source of the river was not important in its own right, as is perhaps obvious, to some British officials. But to the British frontier officers and policy makers, the search provided the opportunity to explore the geography of the region. This emphasis on finding the source reflected the intellectual climate of the time, which was marked by the emergence of a "militant geography" that searched for certainty in geographical knowledge.[22] The objectives of colonial geographical explorations and geopolitical strategy were informed by certain modern (rational) sensibilities as they sought to make knowledge and territorial control unambiguous. This is reflected in the overall objective of the exploration and colonial projects: the former looked for a single point, a source, while the latter looked for a single line, a boundary. Writing about the British exploration of the source of the White Nile in the second half of the nineteenth century, the British novelist and biographer Tim Jeal (2012) states that finding the source of rivers became the leitmotif of British exploration. As Edwards Evans-Pritchard, noting the dominant mode of inquiry (epistemology) among Victorian explorers, stated, "There was everywhere a passionate endeavor to discover the origins of everything . . . an endeavor always to explain the nearer by the farther" (1950, 124). Thus, finding the source of the river or of the original Aryan race amounted to getting to the essence of one's object.

The Search for the Original Aryan Race

Linked with the story of a sacred river and its origins, the region was also depicted as the origin of the Aryan race. Thus there was a great deal of speculation about the true identity, or "race," of the various people who lived in the region. The speculations and sometimes debates among British officials about the identity of the people must be understood within the context of prevailing ideas about race.

During the latter half of the nineteenth century, Victorians were immensely interested in the issues of race, climate, and civilization and their effects on one another. Western Europeans in general and English Victorians in particular were interested in finding out how such a huge diversity of cultures and physical types of humans could be explained and what were the natural laws, sequences, and processes through which human history moved. Social evolutionary theory provided the answer. It identified cultures according to a sequence of development—from savagery to barbarianism to civilization—viewed as a universal process applicable to the entire human family. Social evolutionary theory and the conception of race within it were later challenged by Darwin's understanding of biological evolution and his theory of natural selection based on adaptation and succession. If the former placed race within the forces of history, albeit driven by a "natural" law, the latter placed it outside history: the continuity of race was driven by chance and random events (Ingold 1986, 8; Stocking 1968, 97). Perhaps the most defining influence on the concept of race during this period was British explorer and naturalist Alfred Wallace's insight that geographical isolation, remoteness from the general population, resulted, over a long period, in the formation of distinct species or races. (A contemporary of Darwin, Wallace is credited with independently coming up with the theory of evolution based on natural selection.)

The British explorers of the region were motivated not only by the search for the source of the Oxus but also by an interest in ethnology. As they explored mountain passes, measuring their heights and recording the natural features and resources of the area, they also speculated on the true identity of the people of the region. The first British explorer to the northern frontier region of India was William Moorcroft, the superintendent of the stud farm of the East India Company in Calcutta. Moorcroft never traveled to Hunza; rather, he went from Kashmir to Kabul and then to Bokhara. Trained as a surgeon, Moorcroft later became a veterinary surgeon. He was interested in different breeds of domesticated and wild animals that were of economic importance; for example, in the first decade of the nineteenth century he traveled to the Tibetan Plateau in search of the famous Pashmina goat. Moorcroft's later travels to the northern frontier region between 1819 and 1825 were supposedly carried out to find the fabled Central Asian horse and establish trade relations with the rulers of the small states on the northern frontier of what was then the Sikh Empire of Maharaja Ranjit Singh. Moorcroft died of fever on his return journey in 1825, and several of his articles were published posthumously.

A book based on his original travel accounts was published in 1841, edited by Horace Hyman Wilson, English Orientalist, assistant surgeon general of the Bengal Presidency, member of the Royal Asiatic Society of Bengal,[23] and employee of the East India Company. According to Wilson, Moorcroft's travels contributed greatly to the rising science of geography, which was illuminating hitherto dark and unknown areas for the benefit of Europeans. Moorcroft's book is full of descriptive accounts of geographical features, natural resources, domestic production, political organization, and social history. Although Moorcroft clearly sees local people as less civilized than Europeans, he does not portray them as the radical "other." He shows a mild admiration for the local culture and the extent of its sophistication, while also clearly showing interest in information on commerce, communication, ethnologies, and political boundaries.

Moorcroft's work resembles those of Orientalist scholars in general before the era of the Raj; they helped produce sympathetic images for the British of Indian society. The overriding concern of these scholars seems to be trade with Central Asia, especially with China, rather than the defense of India from the Russian threat. It is for this reason that Christopher Bayly has stated that Moorcroft's journal reveals "a mind poised on the cusp between the classical Orientalism of Sir William Jones's generation and the statistical empiricists and empire-builders of the generation of Everest and Burnes" (1999, 138).

Moorcroft's book provides sketchy accounts of the geographical location and society of Hunza, or "Hounz": "Beyond Nagar, and nearer to the Pamer mountains, is the district of Hounz, also inhabited by Dungars. Kanjut is the name of the capital, which is the residence of Selim Shah. From this province a difficult pass leads across the mountains to Badakshan" (1841, 266). It is not clear who the Dungars were, and there is no mention of that name in any later references to the people of the region in colonial accounts. It was Moorcroft's Indian assistant who first used the term Dard for the local people, but it became popularized by Moorcroft's account (Clark 1977, 330). Moorcroft identified the people surrounding the Gilgit region as Dardus. This included as Dards the people of Gilgit, Astor, Chilas, Darel, Tangir, Punyal, Ishkoman, Yasin, and Chitral. Moorcroft wrote that he also thought that the Baltis, from neighboring Baltistan, had descended from the Greek general Alexander the Great, but he said that he could not find any evidence for the veracity of this claim. While Moorcroft does not provide any more information about the Dards, he seems to have believed, without stating it explicitly, that they were of the original Aryan race. In a footnote, Wilson the editor writes of the Dards: "Few people can be traced so for long a period in the same place, as they are evidently the Dardas of Sanscrit Geography, and Dradae, or Daradrea of Strabo" (Moorcroft 1841, 266).

The mention of Dards in Sanskrit (Clark 1977, 332) and Greek ancient texts (including the works of Strabo, a Greek geographer, historian, and philosopher born in approximately 64 BC) allowed Wilson to create a new category of "Dards" to refer to the people of the region.[24] Moreover, in his preface to Moorcroft's work, Wilson described the region between India and China and between India and Russia as the "cradle of civilization" (1841, xv). Hunza itself remained unmarked in Moorcroft's account, perhaps because of a lack of its perceived strategic importance to the British at this time.

Godfrey T. Vigne, a gentleman traveler from England, traveled to Baltistan on the south side of the Indus in 1834–5. He published a book based on his travel accounts just a year after Moorcroft's. Vigne gives a descriptive account of the area's geography and natural features, such as rivers, glaciers, and mountain passes. Although he never visited Hunza, Vigne speculated on the identity of the Baltis, who had generally been considered Tibetan and hence a different "race" of people from the Aryans. Referring to the name of the Balti raja, "Gylpho" in the Balti language, Vigne states that it was the same as "Guelph," Danish for *king* (1842, 233). Vigne is less damning of the locals than other explorers of his time, who saw them as isolated and lacking any history. Vigne tells tales of Baltistan's history, of local rajas and their small states built on control of human labor for the colonization of land through irrigation and herding. He writes admiringly of the raja of Skardu, Ali Sher Anchan, who in the sixteenth century expanded the boundaries of the state up to the Afghan border at Chitral in the west and up to Leh near the Tibetan Empire (254).

Between 1845 and 1846, the British defeated the Sikh Empire of Lahore and sold its northern district of Kashmir to the Maharaja Golab Singh under the Treaty of Amritsar in 1846.[25] Through this treaty, the entire mountainous region of Ladakh, Baltistan, Astor, and Gilgit, which had been subjugated by the Sikhs in the first half of the nineteenth century, became an appendage to the lands of the maharaja of Kashmir. The fate of Hunza remained unclear; while the British thought that Hunza was not a territorial part of the Kashmir state, the maharaja thought otherwise.

The first attempt by the British to make diplomatic contact with Hunza occurred in 1847, when British lieutenants Patrick Alexander Vans-Agnew and Ralph Young were sent there. We know, however, that Vans-Agnew and Young were refused entry to the valley, which triggered a long spell of fitful hostilities between the government of India and the rulers of Hunza. Their report was never made public and was believed to have been lost when Vans-Agnew was murdered (by the deposed Sikh governor of Multan, Mulraj, after the Sikhs lost the city to the invading British Army) in Multan after his return in 1847 (Drew 1875, 440; Biddulph 1880, 8). But Dorothy Woodman (1969, 354) has reported that in an enclosure sent with secret letters from India, Vans-Agnew made some remarks with respect to the position of Kashmir's northern boundary and suggested the Karakorum watershed as the natural boundary between South and Central Asia.

Between 1840 and 1860, especially after the first Afghan war, the Russian threat subsided, temporarily halting the forays of explorers into the region. Meanwhile, the area came to be seen as more dangerous: the Kashmir government of Maharaja Golab Singh had fought two wars against the combined forces of Hunza and Nagar[26] and had been badly routed (Drew 1875: Durand 1900). The lull in exploration was short lived. The fall of the khanates of Khiva, Merv, and Bokhara to the expanding Russian Empire from the east in the late 1860s and 1870s ushered in a new wave of explorers into the northern frontier region of India. During the 1860s, fresh reports of a Russian menace had gained popularity in the local British press. Trench wrote in 1869, "To anyone who has, during the last eighteen months, watched the tone taken by the leading organs of the press, and therefore adopted by public opinion in England, it will, I think, have been perceived that danger that the Russophobists have always been foretelling is a real one after all . . . which will require to be faced far sooner than had been commonly supposed" (29).

In 1864, the government of India sent Gottlieb Wilhelm Leitner to head the newly established Government College in Lahore. Leitner was an expert in Oriental languages; by the time he arrived in Lahore at the age of twenty-three, he had already mastered over fifty languages and worked as a professor of Arabic and Muslim law at King's College London. By this time the rising influence of utilitarianism had created a backlash against those early scholars of the East India Company who advocated taking seriously Indian arts, literature, and sciences. But Leitner was an oddball. At Government College, which later became the University of Punjab, he fought hard with the British authorities and managed to include in the curriculum classical Oriental languages such as Persian, Sanskrit, and Arabic. Leitner visited Gilgit in 1866 on the recommendation of none other than the veteran Orientalist scholar H. H. Wilson. He went there to write an ethnology of the Dards, the people first mentioned by Moorcroft on his mission in the 1820s in the book that Wilson edited.

Like Wilson and Moorcroft, Leitner argued that the Dards were supposed descendants of the "Daradas" of Sanskrit literature, but he also identified them as an original Aryan people of great ethnological importance (1889, 49). About the Shins of Gilgit town, he wrote, "I have little hesitation in stating that the pure Shin looks more like a European than any high-caste Brahmin of India" (62). He proposed the name "Dards" to replace "Yaghis" for the people of the region. Leitner postulated that they were neither Indian, Tibetan, nor European, so they must represent a different group—the original race of people of the region. Moreover, reference to them in Greek texts suggested that they predated Alexander's march to the east and therefore they must be the common ancestor of Indo-Europeans. He believed that these Dards had preserved their racial identity because of the high mountain barriers of the regions in which they lived; their geographical remoteness had stopped intervention from the outside. Based on that, Leitner worked hard to introduce new categories of the people and languages of the region that is now Hunza and the area south of it, which he called Dardistan. Unlike Frederic Drew (1875) and John Biddulph (1880) later on, Leitner included the people of Hunza as Dards as well. About twenty years later, Biddulph (8) criticized Leitner (whose work had been published in the 1860s in English newspapers and in the *Orientalist*, the journal of the Royal Asiatic Society) for conjuring up a new race called Dards. He stated that the region was a mix of many "races" and ethnicities that couldn't be categorized as one overarching identity. He also thought, like Robert Shaw (1871), that the Dards were not the original inhabitants of the region but rather had come from the northwest.

Some argue that the British speculation that the region was occupied by a single race should be seen as an attempt to imagine and administer the area as one political unit (Allan 1993, 26). But it is not clear if this is true. Biddulph's criticism of Leitner was unjustified, and given that Leitner objected to the forward policy of hawks like Biddulph, the criticism could have stemmed from professional rivalry. Leitner was not naïve enough to think that this vast region between Kashmir and the Central Asian watershed was inhabited by one common race; he said that they might be called Dards for the sake of convenience. This explanation seems to align with Graham Clark's (1977, 342) argument that the name Dards, and hence the ethnicity, was invented by nineteenth-century philologists, colonial officials, and travelers for the purpose of political convenience and simplification of classification.[27]

The Dards were, for Leitner, an example of people living close to nature and natural conditions, and he believed that a glimpse into their lives and language might reveal how today's "civilized" people once lived and spoke. Leitner argued that these communities should be left alone, that their freedom should be preserved to prevent them from degenerating like the Aryan races who had settled further south in Kashmir and northern India.

Exploration of Scientific Race

By the second half of the nineteenth century, a major intellectual transformation had taken place in the way in which the British represented and wrote about the people and landscape of the northern frontier region.

Though the accounts of explorers and travelers were still informed by the classical ideas of a search for an original Aryan race, they were now also infused with the rising influence of scientific and social Darwinism. Through Darwin's association with the RGS, these ideas percolated into the writings of many RGS-funded explorers. Darwin was a member of the society, which funded many of his research and field trips, and he regularly wrote in the society's journal and participated in its members' meetings (Stafford 1989, 9, 22). One of the main ideas behind Darwin's theory of evolution (as shown earlier, it was actually Wallace's) was that insurmountable barriers in landscapes created isolated societies that over time gave rise to distinct races. For example, Alfred Wallace writes in his famous works *The Malay Archipelago*: "We are . . . led irresistibly to the conclusion that changes of species, still more of generic and family form, is a matter of time. But time may have led to a change of species in one country, while in another the forms have been more permanent, or the change may have gone on at an equal rate but in a different manner in both. In either case the amount of individuality in the production of a district, will be to some extent a measure of the time that district has been isolated from those that surround it" (2000, 217).

When George Hayward was sent to collect political and geographical information about the small states of Hunza and Gilgit in 1870, he believed that the people of the region belonged to a Dardic race, but he fell short of calling them an original people or an original Aryan race that was the ancestors of Indo-European people. Like Leitner, he also included the people of Hunza as Dardic. More important, Hayward thought that the physical barriers in landscapes, such as rivers and mountain ranges, coincided perfectly with racial boundaries. "The inhabitants of Dardistan, in which may be included Gilgit, Chilas, Hunza-Nagar, Dilail, and Upper Chitral, are a fine, good-looking, athletic race, and the difference of race is *at once* perceived on crossing the Indus. Light and dark brown hair, with grey, hazel, and often blue eyes, are seen" (1871, 3; emphasis added).

This represents a subtle change from earlier accounts in the British representation of the region. Descriptions of physical boundaries, such as rivers and mountain ranges, were discussed in earlier accounts, too, but by the 1870s, these boundaries had acquired a new significance. It is doubtful that the change in racial character was as sudden as Hayward actually described. The implication of Hayward's account for the cultural representation of the local society was that the natural barriers in the physical landscape rendered the people living there distinct and discrete races, bounded in small and impenetrable spaces. These spaces acted as nature's mechanism for retaining races in their original conditions, freezing them in time.

Curzon, who on the one hand was enamored by the story of Mount Meru and a primeval Aryan paradise, maintained on the other hand that physical barriers in landscape can give rise to distinct races. About the people of Hunza, he emphasizes the impenetrability and isolation of these societies based on their immense mountain geography. "People who are embedded in this remote and ancient ethnological stratum have retained an *individuality* of their own for centuries" (1926, 145). While Curzon may have speculated on the status of the Darads, an original Aryan race, he also seems to have held the view that the physical separation of these societies due to barriers in the landscape was producing distinct races. It seems that Curzon also believed the people of Gilgit and Hunza were the same (Dardic) people, as he writes: "South of the Burzil the people appeared to belong to the same race as in the Kashmir plains; but northward we encountered a new type, of which we afterwards met with numerous *specimens* as we proceeded to Astor and Gilgit and passed on to Hunza and Nagar" (169; emphasis added).

In 1889, 1890, and 1891, Francis Younghusband visited the region on a special assignment to win the allegiance of the local rulers against the advancing Russians. Younghusband reiterates ideas of remoteness and isolation due to landscape barriers and of preservation of races in their original form in this passage about the people of Yarkand in Chinese Turkestan: "Their mountain barriers shield them from severe outside competition, and they lead a careless, easy, apathetic existence. Nothing disturbs them. Revolutions have occurred but they have mostly been carried out by foreigners, such people are, as might naturally be inferred, not a fighting race. They are a race of cultivators and small shopkeepers, and nothing more, and nothing would make them anything more. It is their destiny, shut away here from the rest of the world, to lead a dull spiritless, but easy and perhaps happy life, which they allow nothing to disturb" (1896, 190–91).

Whether used as a means of explaining the distinction between different mountain groups as different races or as a means of underlining the preservation of individual races through isolation, this emphasis on geographical remoteness and lack of familiarity with the region reified the cultural difference—and reinforced the physical distance—between the British and the people of the region. What had started in the early part of the nineteenth century, under the influence of Orientalist narrative, as a search for a common origin of Indian and European peoples in remote time and space had by the end of the nineteenth century, in the light of scientific Darwinism, ended with justification for the differences between the two peoples.

Conclusion

If remote places in Orientalist narrative, colored by a biblical notion of time, harbored ancient kin, in Darwin's evolutionary discourse, informed by natural time, they harbored distant kin. According to the Orientalist narrative, what was remote in space and time had been near in sociological space, but according to the Darwinian narrative, what was remote in space had become distant in both time *and* in sociological space. Thus, remoteness did not lose its significance with the rise of social Darwinian theory; rather, it found a new meaning. The remote place was not only a primeval paradise where an original people lived but also a geographical space where one could observe evolutionary processes as the act of origination. The formulation of this meaning of remoteness was also part of the intellectual development of the discipline of natural history, in which remote places were seen as where nature and its processes reign.

Michael Dove (2011, 250), commenting on the spice trade between Asia and Europe in the early modern period, writes, "What was distant, unknown, and frightening drove this commodity trade. If it was proximate, known and familiar—like some of Europe's home-grown spices . . .—it was not as valuable or desirable." In the same fashion, the journeys to the source of a sacred river and the original Aryan race drew their value and credibility from the remote locations in which they were conducted. Accounts of these journeys whetted the desire for adventure and appealed to the imagination of the general public, fellow travelers, officials, and traders of the British Empire. But these journeys also helped the British find the boundary of their empire. Once the British had lifted the veil from the region's mysteries and armed themselves with a new ideology of radical cultural difference, they looked to control the Karakorum watershed on the frontier with Russia, and they planned to govern this remote region. But in order to do that, they had first to gain access to it.

CHAPTER 2

The Friction and Rhetoric of Distance and the Alterity of Hunza

Peppered throughout the literature produced about the northern frontier region of British India during the nineteenth and twentieth centuries, we notice statements about the region as a civilizationless area. In 1889, when Francis Younghusband traveled north from Srinagar in Kashmir to Kashghar in Chinese Central Asia through Hunza territory, he wrote this about his travel: "Kashghar was well known, too, from the Indian side, and there was a Russian consul stationed there. So when I reached the place I appeared to have arrived again on the fringes of civilization" (1896, 64). When Younghusband returned from Kashghar and entered Kashmir, he wrote, "My first care . . . was to go off to one of the native shops . . . and purchase a clean shirt and to have my hair cut, my beard shaved off, and to get a good wash. . . . I had expended nearly two hours upon these preparations for my plunge into civilization" (211). In 1898, Jane Duncan, one of the few women adventure travelers to the region, writes in the same vein: "One of the joys of expedition was getting away from dress with its worries as distinguished from mere clothes, and many a time after returning to civilization I longed to be in the desert again where crows and goats did not care what I wore" (1906, 11).

These statements of coming out of and into civilization were not merely rhetorical. From the way they were used, especially with reference to a physical space, it is clear that for these travelers, civilization, in addition to having a sociological or temporal dimension, also had a spatial dimension. The British imperialists imagined the vastness of their empire as a sea of civilization with remote islands of savagery and primitiveness. For British colonial officers and travelers, the mountains of the region were barriers to the movement of civilization into communities that, therefore, existed insulated from its influence. For example, John Biddulph, first political agent of the Gilgit Agency, wrote, "Aided by nature in preserving their independence, and partially isolated from one another, the people of the country have formed themselves into a number of separate communities which have existed for generations within the same narrow limits. Living the same life and following the same customs as their forefathers did hundreds of years ago, they have remained unaffected by the changes that have taken place around them" (1880, 2).

The topographical conditions that put impediments to fast travel in the region surrounding Hunza put it in a category of "primitive" places. The British explorers, sportsmen, merchants, spies, officials, and others who came here all experienced what David Harvey has called the friction of distance, defined as the time or cost expended to traverse a particular distance (2006, 122). James C. Scott, in his study of the social and physical aspects of resistance, has further refined the term to show that different "terrains" offer different levels of resistance to mobility or travel, and hence to outside interference, particularly by centralized states. For example, it is much easier to travel over water courses than over a mountainous terrain (2010, 47).[1] Within the context of the Great Game, the most important goal for the British in this region was to construct a mule track on which they could move an army and supplies from Sri Nagar in Kashmir all the way up to the Russian and Chinese frontiers through Hunza territory. That Hunza's remoteness was a product of the friction of distance in the region is reflected vividly in the accounts of British officers and travelers who surveyed and supervised the building of roads in the region between the 1870s and 1890s.

Hunza's remoteness emerges not only in overcoming the friction of distance through distance-busting technologies but also in the rhetoric of distance. Michel de Certeau has argued that European colonial travel writings resulted in this rhetoric of distance, which rendered all distant societies radically different in terms of their cultural logic and material practices. "First comes the outbound journey: the search for the strange, which is presumed to be different [by] the place, assigned it in the beginning by the discourse of culture. This *a priori* of difference, the postulate of the voyage, results in a rhetoric of distance in travel accounts. It is illustrated by a series of surprises and intervals (monsters, storms, lapses of time), which at the same time substantiate the alterity of the savage, and empower the text to speak from elsewhere and command belief" (1986, 69).

While the friction of distance in the region, and hence Hunza's physical remoteness, was inevitable given the shape of topography and terrain, the rhetoric of distance produced Hunza as a remote area in a different way. According to it, what was geographically distant was also sociologically distant, that is, unfamiliar and exotic but also dangerous and threatening.[2] British accounts of the people of Hunza in the preconquest phase of 1891 show them as sociologically distant others who were beyond reason.[3] For example, Younghusband (1896) and Knight (1900) reported in their accounts that Hunza had an outlandish cultural tradition of presenting their wives to guests as sexual gifts.[4] Durand (1900) narrated a story in which Mir Safdar Ali challenged one of his Gurkha marksmen to shoot at a distant Hunza man to demonstrate his marksmanship. But perhaps the most long-lasting rhetoric of distance about Hunza was its alleged caravan-raiding practices, which in de Certeau's words above "substantiate[d] the alterity of the savage[s]."[5]

In the colonial rhetoric of distance, the people of Hunza were imagined as eccentric and beyond reason because of their remoteness and isolation. The mir is presented as a rather whimsical and ignorant character who thinks, for example, that the "Empress of India, the Czar of Russia, and the Emperor of China were chiefs of neighboring tribes" (Younghusband 1896, 285). The mir was perhaps employing deliberate rhetorical and symbolic figures in his political discourse, which were at times misunderstood by the British. Rather than trying to understand the subtext of such remarks in Safdar Ali's responses, the British presented them as exemplars of the mir's unreasonableness and hence otherness (Hussain 2006). While the rhetoric of distance constructed Hunza as a remote place in the accounts of the British, we will see that the ruler of Hunza did not see himself as such. Rather, he saw himself at the *center* of the three greatest empires on earth.

Literature on colonial sociology (Cohn 1996; Skaria 2001) and European conceptions of non-Europeans (Fabian 1983; McGrane 1989) shows that distance and ease of travel were important criteria in colonial/European categorization of space and of the people therein. So, generally speaking, to the British what was distant and inaccessible was also deemed low on the evolutionary scale. Ajay Skaria has described a similar sentiment among colonial officers toward the hilly tracts of the Western Ghats. Harsh landscape that hindered movement was considered part of the habitus of the people who lived there. They were, as a result, considered intractable and intransigent (Skaria 2001, 54; Pandian 2008). Against stiff resistance from the rulers of Hunza to any road-building projects, the British articulated this as a cultural project: to bring civilization to Hunza or Hunza to civilization.[6]

The Friction of Distance

During the last three decades of the nineteenth century, the remote regions surrounding Hunza were connected to the British Empire through various road-building and improvement projects.[7] The main objective was to connect Srinagar with the Russian frontier on the Karakorum watershed. The stupendous mountainous terrain made going slow and hazardous, but once the road had been built, actual military campaigns were not much problem for the British because of their better discipline, training, and weapons. The pace and enthusiasm with which road-building projects were implemented reflected the real and perceived threat of Russian invasion, the popularity of the forward policy narrative in British political circles, and the personal character of many frontier officers during this period.

Prior to the British arrival in the region, the Kashmir state had had its own version of the "forward policy." In 1842, when Kashmir was still part of the Sikh Empire of Ranjit Singh of Punjab, the governor of Kashmir, Golab Singh, had conquered Gilgit, a small principality about sixty-five miles south of Baltit, the capital of Hunza, and set up a district administrative structure. Between 1848 and 1869, Kashmiri forces, on behalf of the British, made three attempts to subjugate Hunza to the north, but on each occasion they were routed (Drew 1875, 435–43). Throughout this period, Ghazan Khan, then the mir of Hunza, politely but firmly refused to allow any posse of British or Kashmiri officers to pass through Hunza territory. In the 1850s, the Kashmir forces had occupied the fort of Chaprote on the southern boundary of Hunza and in the neighboring state of Nagar. Both Hunza and Nagar had made repeated attacks on the Chaprote Fort, which they considered to be their property. Nevertheless, in 1870, Ghazan Khan recognized the suzerainty of Kashmir and started paying its ruler an annual tribute in return for a handsome subsidy (457). The hostilities between Hunza and Kashmir stopped, at least for the moment.

The lull in hostilities spurred British forays into Hunza and the surrounding regions and started perhaps one of the most intriguing periods of the Great Game. As mentioned in the last chapter, in 1872 and 1874, the

British and Russians had agreed to demarcate their relative spheres of influence at the source of the river Oxus without actually knowing exactly where it was located. In other words, the actual border had not been formally delimited and demarcated, and thus this entire frontier zone was up for grabs by the imperial powers. British anxieties about Russian threats were not helped by the unfriendly and obstructionist behavior of the maharaja of Kashmir, Ranbir Singh, and his officials, upon whom the British had depended for their travels in the region. The British desperately wanted to take administrative control of the region.

In 1875, John Biddulph went to the region to collect intelligence about Russian movements and wrote a report and later a book, *The Tribes of Hindoo Koosh*. His is the first systematic account of local tribes, and his book became a standard reference text for most British officers. Biddulph was definitely taken by the grandeur of the mountains and their sheer size. Commenting on the lack of knowledge about the region due to the difficult mountain terrain, Biddulph states, "The apparent neglect has been caused by the almost inaccessible nature of the country. In no other part of the world, probably, is there to be found such a large number of lofty mountains within so confined a space" (1880, 1).

A practical-minded person, he thought of communication systems in the region in terms of access, routes, supply lines, weather conditions, and so on. About the topography of the region, with an eye on accessibility, he wrote: "This immense mass of mountains is intersected by numerous deep valleys, and these, owing to some peculiar geological formation, which I have not remarked in other parts of the Himalayas, are generally narrower at their mouths than higher up. It is not unusual to see among them valleys of from 10 to 30 miles in length, supporting a population varying from 500 to 5,000 souls with an embouchure so narrow that it is difficult to find a pathway beside the torrent which issues between overhanging rocks" (1880, 1).[8]

Returning from his travels in 1875, Biddulph wrote a report reflecting his somewhat alarmist assessment of the Russian threat. He recommended that Britain extend the frontier of the empire to the northern border with Hunza. And he noted that this should be accomplished by force, if need be, obviously mindful of the expected resistance to this plan by the rulers of Hunza, thus setting the tone for British "forward policy" in the region. Biddulph overplayed the ease with which an invader from the northern passes of Kilik, Mintaka, and Khunjerab at the Karakorum watershed and on the border of Hunza could bring down a large and well-equipped army. The return to power in 1875 of Benjamin Disraeli, a forward policy advocate of the Great Game, provided momentum to expansionist impulses in British Indian policy circles. Two years later, the British convinced the maharaja of Kashmir to let them establish the Gilgit Agency, and Biddulph was appointed its first political agent. The British now had a working political administration in the region with which they engaged the local tribes.

Biddulph saw the expansion of British administrative and military control to the Hunza frontier as his foremost objective. He put great importance on strengthening the defenses of Chaprote and Chalt villages, north of Gilgit town and within a few miles south of the Hunza border. Although the independent states of Hunza and Nagar both claimed the area as their territories, Biddulph considered them to be within the boundaries of Gilgit and hence part of Kashmir. Indeed, a scanty Kashmir garrison occupied two strategic forts, one at Chaprote and the other at Chalt farther to the north. Biddulph urged the maharaja of Kashmir to increase forces at the forts and assert Kashmir's authority, but the maharaja refused.[9] In 1880, there was a rumor that combined Hunza and Nagar forces would make a bid to wrest control of the Chaprote Fort from the Kashmir forces. This allowed Biddulph to make the case for enlarging the size of the Kashmir garrison, which was eventually increased from 50 men to 250. The agency was abandoned in 1881, three years later, when the liberals came back to power, and also because Biddulph was unable to convince the government in Calcutta and London that the Russian threat was credible or that Hunza was in contact with it.

The Russian threat resurfaced in 1885, when the Russians seized Afghan territory near an oasis at Panjdeh, south of the Oxus River in the Pamir Mountains. In doing so, they breached the southern boundary of the Russian Empire as agreed upon in 1872. According to the treaty, the territory belonged to the buffer zone of the Afghan Wakhan corridor (Todd 1951). The incident created a diplomatic crisis between Russia and Great Britain. A year later, Safdar Ali, the youngest son of the mir of Hunza, Ghazan Khan, murdered his father and became the mir.[10] Two years later, in 1888, he combined his forces with the Nagaris and sacked the Kashmir-controlled Chaprote Fort for the second time in a decade.[11] At this point, Lord Dufferin, viceroy of India, wrote to London stating that Hunza, though a small state of not much military strength, held a key strategic position in the defense of India by holding the mouth of the passes to the north. Lord Dufferin argued that Hunza was too important not to be ruled directly by the British. In 1889 the Gilgit Agency was reestablished, with Captain Algernon Durand, brother of Mortimer Durand, secretary of state for India and the architect of the famous Durand Line—a border that divided Afghanistan and what was then British India (today, Pakistan)—that would be established six years later, was sent by the government as the political agent.[12]

Durand was a decidedly ambitious man. His book, *The Making of a Frontier*, expresses in classic form the muscular nature of British imperialism of the time. Durand's job was the same Biddulph had held a decade ago: to extend British territorial control to the foot of the passes of the Karakorum watershed that divided Central and South Asia. That meant constructing a supply route from Srinagar to Gilgit, as permanent occupation required the presence of a sizeable army in Gilgit. This army had to be fed, and people in the region already lacked grain to eat. Writing about his first year as the political agent in Gilgit, Durand recalls, "My time was now fully occupied at Gilgit. The year before I had drawn up complete schemes for putting postal and transport arrangement on a proper footing" (1900, 230). During his tenure, Durand rapidly opened up the region by building roads and other communication infrastructure, such as telegraph and mail systems, while maintaining a constant dialogue with the mir.

The task that Durand set out to accomplish during his tenure was improving—widening and extending—the main route linking Kashmir (and the rest of India) with Gilgit and Hunza. When Durand took over the administration of Gilgit, the route was a two-hundred-mile-long dirt track, which was suitable for pack animals and horses for a little more than half of the length; the rest was passable only on foot, ascending to almost fourteen thousand feet where it crossed the Burzil (see figure 3).[13] The entire trek lasted twenty-one days. Then, because of the lack of any major passes on the route, it took only three more days to travel the sixty-five miles to Baltit, the capital of Hunza, north of Gilgit. Francis Younghusband, who traveled at breakneck speed from Baltit to Gilgit in 1889, wrote this about his experience: "Two days from Baltit—after passing over sixty-five miles of most execrable roads, by paths climbing high up the mountain-sides to round cliffs or pass over rocks and boulders, and by galleries along the face of a precipice—I reached Gilgit" (1896, 288).[14] Most tracks ran alongside a river but often involved long detours, climbing up and down the mountainside to avoid numerous cul-de-sacs. Thus, travel in this region was measured in terms of number of days rather than actual distance, because it could not be assumed that the topography, and thus the pace of travel, would remain uniform throughout the journey.

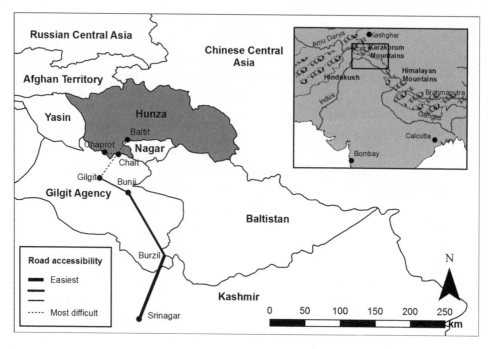

Figure 3: Friction of distance between Kashmir and Hunza (map drawn by Chris Milan)

When Durand took control of the Gilgit Agency, the road was poorly maintained. Durand was fully aware of the advantage of the new mode of transportation, mule carts, over the existing system, the use of porters. Before direct British involvement in the region, the Kashmir state had required local people to provide free labor, *begar*, to maintain the road and provide transport to state officials traveling in the region.[15] But using humans as beasts of burden was inefficient compared to using mules. Durand states: "Over bad roads coolies can only carry a load of from sixty to eighty pounds, in addition to five days of food for themselves. After the first few days they begin to consume what they carry, and bad roads make short marches. Mules carry four to five times as much as men, and only eat twice as much. Roughly stated, that is the whole problem of supply over frontier roads" (1900, 232).

In 1890, the road north from Srinagar to Gilgit was suitable for a pony or mule track up to Astor, some eighty miles short of Gilgit. In the summer, before the Burzil or Kamri passes closed for the winter, Durand needed to move some 360 tons of grain to feed the thousand-strong army. His troops comprised locally raised levies (Kashmir State) and imperial (British) forces.

Durand tried to increase food supply locally by building water channels at Gilgit to bring barren land under cultivation for grain production for the troops (1900, 232). At the same time, at Gilgit, Durand tried to improve the existing road north to the fort of Nomal, which was eighteen miles from Gilgit and had been besieged by the people of Hunza and Nagar three years before.

In the summer of 1890, Durand wanted to move his troops to Nomal to reinforce the Chalt Fort on the boundary of Hunza and Nagar further north as there were rumors that a combined force of Hunza and Nagar fighters was about to assault the fort. Chalt fort, which lay another twelve miles north of Nomal and was manned by a small party of Kashmiri troops, was under constant threat of invasion from Hunza. Durand's first goal was to open up the road to Nomal. Describing its condition, Durand wrote, "For fourteen miles the road was fairly practicable for mules, but for four miles it was impassable in summer" (1900, 232). He instructed his sappers to complete the road as quickly as possible, but the work was tough on such unstable terrain. "The last hundred yards of the worst bit of the cliff nearly broke our hearts. The hillside was formed of alternate layers of sand and loose water-worn stones, and, cut back as we worked, we could not get a firm ground" (233). But the sappers of the Gilgit Agency succeeded by early 1891 in building a mule track six feet wide to Nomal Fort.

In the spring of 1891, Durand wanted to move troops from Nomal north to Chalt Fort, the last hold of the Kashmir forces at the Hunza frontier, but again the road was in abominable condition and he had practically to build a route from scratch. "The last part of the road had been so bad that I could not bring on the guns, but I turned out every available man of the Nomal garrison and set them to work to improve it, and arranged for the guns to be brought later" (1900, 242).

The people of Hunza were fully cognizant of the fact that their natural advantage of remoteness based on the "friction of distance" in hostile situations was being taken away. Access to their region by improved technology of transportation meant that they were vulnerable to outside forces. In May 1891, the people of Hunza and Nagar sent a warning to Durand not to continue his project of road building into their territory, even demanding that he vacate Chalt Fort, which the "tribesmen looked at as the strings of their wives' pyjamas" (Durand 1900, 242). They announced that any road construction to Chalt would be considered an act of war. Durand was certain of an imminent attack. "As the scraps of information from all sides reached me, and were collated and pondered over, it seemed to me that all tended to show that real mischief was brewing. The threads I held in my hands persistently wove themselves into one pattern, and that pattern meant war" (235). Justifying his action as defense against the impending attack, he moved troops closer to Hunza's border by extending the road to Chalt Fort to reinforce the precarious control of the fort. The people of Hunza saw this as an escalation in hostilities. They refrained from attacking the fort, but they did refuse to pass through their territory letters from the Indian government to the British resident in Chinese Turkestan. The British declared this a violation of the 1889 agreement between Hunza and British India. Curzon recalled later, "It was in these circumstances that, Colonel Durand's scanty force at Gilgit having been reinforced by officers, men, and guns from India, it was decided to send an ultimatum to the recalcitrant rulers, informing them that a new fort was to be built at Chalt and that a military road would be constructed to Hunza on one side of the river . . . so as to give freedom of access to the frontier, which the Indian Government had determined to hold" (1926, 187).

In October 1891, with the roads from Gilgit to Nomal and from Nomal to Chalt complete, guns, grain, and soldiers were ready to move into Hunza. Durand had sent an ultimatum to the rulers of Hunza and Nagar not to resist British troops entering their territory to build the road to the Russian frontier. "If the Chiefs refused their consent to the road-making they were to be told that the roads must be made and unless they complied with our demands, troops would enter their country and make the roads in spite of them" (1900, 246). When the ultimatum was rejected in December 1891, British forces invaded Hunza—in a matter of three weeks they had taken over the state. Before the British arrived, Safdar Ali fled to Kashghar in China, where he was received as an official guest.

Writing about the Hunza campaign, twentieth-century historian John Keay wrote, "Those who described the war as one of the most brilliant little campaigns in military history were thinking as much of the logistical achievement as of the heroics at the front" (1996, 491). With its conquest, Hunza became the forward observation post for monitoring the movements of the Russian Army. From an unknown frontier, Hunza had become a remote frontier of the British Empire whose importance waxed and waned with the geopolitics surrounding the Great Game. In the end, the road-building projects that overcame the region's friction of distance introduced civilization through violence and military means. Not all in the British administration thought this method was the most appropriate. Archival evidence suggests that the resident in Kashmir and the secretary of state did not fully share Durand's views. They also did not support his eventual invasion of Hunza because they thought that when a road was built in a region, its rulers would naturally accept British rule.

The Rhetoric of Distance

One of the most enduring cultural representations of the people of Hunza to this day is that they were once caravan raiders. They are featured in colonial travel literature as raiders and plunderers of Kirghiz encampments and of trade caravans that traveled between India and the Chinese Central Asian towns of Kashghar and Yarkand. They are often described in colonial literature as uncivilized savages and "wild looking men" (Younghusband 1896, 260). The region in general was referred to as Yaghistan[16] (Yule 1872, xxvi; Biddulph 1880, 4), meaning, according to Durand, "land of the ungovernable, of savages" (1900, 177). The British commonly used charges of caravan raiding against local people to justify territorial expansion and colonization. Raiding was presented by the British as a sign of the primitive cultural and mental disposition of local people, which is why charges of raiding in British imperial discourse appear side by side with the civilizational discourse of the "white man's burden."

According to anthropological literature, incidences of raiding can be political and administrative in nature. That is, raiding can be seen as a form of tax collection by an authority from its subjects. Perhaps British colonial administrators also knew about this dimension through their experience ruling pastoral societies in Africa and Asia. Both tribes and states raided cattle to collect taxes (Evans-Pritchard 1940, 48). The people of Hunza raided the Kirghiz camps for livestock as a tax, seldom raiding trade caravans, but in the British rhetoric of distance the practice was misrepresented as caravan raiding, thus obfuscating its political economy dimension.

We see time and time again in British travel, exploration, and official literature, especially from the nineteenth century, that hostile pastoral communities are described as caravan raiders, especially people associated with a nonsedentary economic system (Meeker 1980; Salzman 2004). The accusation was particularly powerful as real caravan raiding would have been anathema to the British because of its perceived threat to the legitimate trade that formed the justification for empire in the second half of the nineteenth century. It could also be used as a legitimate excuse to wage war upon a people.

The first few times that caravan raiding by the people of Hunza was mentioned in British literature in the nineteenth century, fifteen years before the incursion into Hunza, it was presented as being a thing of the past. For example, Frederic Drew, who served in the court of the maharaja of Kashmir, wrote in 1875, "In following this road [Skardu-Yarkand] there was formerly—and may be even now—danger from the Hunza robbers, who, issuing from their own country and crossing the watershed by an easier Pass, used to attack the caravans where the two roads met on the farther side of the range" (371). Biddulph, the first political agent of the Gilgit Agency, wrote in 1880 that the last raid may have been carried out in 1865, fifteen years before:

> A fixed subsidy was paid by the Chinese to the Tham of Hunza, who in return gave a nominal allegiance. Under these circumstances, the caravans between Yarkand and Leh were regularly plundered in the valley of the Yarkand River near Koolandoodi by the Hunza people whilst the Chinese authority winked at the proceeding which they were unable to prevent or punish . . . the last exploit of this kind, and the most successful recorded took place in 1865, when no less than 50 camels and 500 ponies laden with merchandise were driven from Koolanooldi to Hunza by way of Ujadbai without opposition. Kanjuti eyes still glisten when they talk of that day, but the establishment of the firm rule of the Atalik in Kashghar put a stop to future proceedings of the kind. On the reestablishment of the Chinese power in Kashghar in 1878, Ghazan Khan [ruler of Hunza] preferred a formal request to me that he might revive his ancient right of "striking the road." (28–29)

But as the pressure on Hunza increased with the establishment of the Gilgit Agency under Durand, its people's reputation as caravan raiders also gained traction. Younghusband referred to the present in describing Hunza's caravan raiding. "In the autumn of 1888 these robbers had made an unusually daring attack upon a large caravan, and had carried off a number of Kirghiz to Shahidula, on the Yarkand road" (1896, 215). He also claimed shadowy sovereignty over the Kirghiz, whom the British eventually used to invade Hunza as a punishment for raiding them. When Younghusband met with Safdar Ali in 1889 he "reminded him [Safdar Ali] that the raids were committed by his subjects upon the subjects of the British Government" (285).

Younghusband uses the word *caravan* in a misleading way, perhaps deliberately. The Kirghiz were pastoralists, and although they may have engaged in various forms of trade on a seasonal basis as part of their migration cycle, "caravan" implies movements by merchants for the purposes of trade. Younghusband and British sources, however, confounded Hunza's raiding of Kirghiz camps with the plundering of trade caravans. It is important to separate out the two "victims" of these raids for historical purposes. Apart from some politically motivated accounts, there is little evidence, contrary to popular belief even today, that the people of Hunza raided trade caravans that traveled between Leh and Yarkand through the nineteen-thousand-foot Karakorum Pass. But there is ample evidence of their raiding on Kirghiz nomads, who grazed animals on either side of the Karakorum watershed.

The Political Economy of Caravan Raiding

The raiding of Kirghiz camps by people from Hunza must be seen in the light of the historical relationship between the Kirghiz and the Hunza state. The Kirghiz are Central Asian nomads who occupied the Pamir valley in Afghan territory and Taghdumbash and Raskam valleys on the border of China and Hunza. They grazed their yaks, sheep, and goats on the northern face of the Karakorum watershed, sometimes crossing the high mountain passes to enter Hunza state to the south. In the seventeenth century, the state of Hunza was involved in efforts to consolidate its political and territorial authority. According to Sidky, during this period, the mir of Hunza, Silim Khan, "decided to bring under his control the areas that lay at the headwaters of the Hunza River occupied at the time by Kirghiz nomads from China's Taghdumbash Pamirs. Silim Khan launched a successful military campaign against the Kirghiz, establishing sovereignty over Shimshal Valley and the Taghdumbash Pamirs. The Kirghiz were obliged to become Silim's vassals and to pay regular tribute to him" (1996, 66). Clearly, the British had overlooked this historical relationship, even though many travellers had mentioned it, and perhaps because of this lacuna they misunderstood the political and economic nature of the Hunza raids.[17]

Small agropastoral states of the region, such as Hunza, attached great importance to grazing areas for collection of taxes and display of their political legitimacy. The rulers of these small states derived their power through control over material wealth such as livestock and their ability to provide protection to their inhabitants from outsiders (Emerson 1996, 100–145).[18] The rulers of Hunza saw the Kirghiz nomads as an administrative challenge because they were not sedentary people, thus making tax collection from them difficult. The Hunza rulers consolidated their control over the Kirghiz by authorizing raids on them when they failed to pay taxes. Moreover, in addition to grazing taxes, the Hunza state also exacted tax on trade caravans that the *Kirghiz* had been raiding.[19]

That the rulers of Hunza viewed raiding the Kirghiz as a way of collecting tax from them is evident in the response Mir Safdar Ali gave to Younghusband when the Englishman "reminded him," in the quotation cited above, that his people were raiders: "Safdar Ali replied, in a most unabashed manner, that he considered he had a perfect right to make raids; that the profit he obtained from them formed his principal revenue, and if the Government of India wished them stopped, they must make up a subsidy for the loss of revenue" (Younghusband 1896, 285). The "raids" that the British complained of were, in this interpretation, merely an assertion by the mir of his right to collect grazing taxes.

All throughout the period being discussed here, the Kirghiz greatly contested Hunza demands of taxes, frequently attacking Hunza graziers and tax collectors. Perhaps it was because of this counterviolence that in some earlier colonial writings the Kirghiz were represented as hereditary practitioners of caravan raiding. For example, Shaw writes, "All we knew was that certain nomads, calling themselves Kirghiz, had formerly rendered the more westerly road to Yarkand unsafe by their depredations" (1871, 104). The Chinese also came to characterize the Kirghiz as a volatile people. During the latter half of the nineteenth century, a part of the territory occupied by the Kirghiz north of Hunza in the valleys of Ferghana and Kashghar fell to the advancing Russians. A large part of the Kirghiz population was displaced to the Chinese-controlled region of Turkestan (Koldys 2002, 223). The Kirghiz's relationship with China was a bloody one. The Chinese saw the Kirghiz as "predators and hard to control" (Lipman 1997, xxx). Around 1850, Hunza helped China squash a Kirghiz rebellion and as a reward received land grants from the Chinese. Most of the grazing land and rights to collect grazing taxes that the rulers of Hunza received were from the Kirghiz-controlled areas on the north of the Karakorum watershed in the Chinese territories of Taghdumbash Pamir and Raskam (Biddulph 1880, 28).

The Kirghiz saw these land grants as illegal; certainly they never acknowledged Hunza's claim of ownership. Thus, Hunza would have to use physical force to impose a grazing tax on the Kirghiz. The British conveniently conceived the physical enforcement of Hunza's policy of collecting grazing taxes from the Kirghiz as "caravan raiding" and "plundering." In light of the geopolitical context and the prevailing rhetoric of distance in accounts written of Hunza, the people were represented as irrational, strange, and beyond the pale. The British represented caravan raiding as a traditional practice rooted in the atavistic hostility of one savage tribe to another. In thus depicting the practice as part of the character of the Hunza people, the British not only obfuscated the political economy dimension of "raiding," they also disguised their own strategic interest in perpetuating the misconception. What is interesting is that both before and after the invasion of Hunza, various British travelers and explorers who went to the region reported that taxes were collected by the rulers of Hunza from the Kirghiz pastoralists who used the pastures at the Taghdumbash Pamir and Raskam valleys (see Shaw 1871; Biddulph 1880; Dunmore 1893). In fact, after the conquest of Hunza, the British negotiated with the Chinese authority on behalf of Hunza regarding a proper procedure for collecting taxes from the Kirghiz.[20]

The Rhetoric of Distance and Mir Safdar Ali

The responses by Safdar Ali, the mir, to British accusations of caravan raiding and their demands for opening up his territory come to us through British accounts. These accounts, as partial and as one-sided as they are, help us to see how the British constructed the image of Safdar Ali as a quintessential other—and how he manipulated the British. In 1890, the British had attempted to sign a treaty with the mir of Hunza, offering him 2,000 rupees per year as a subsidy in return for his promise to stop caravan raiding and to entertain no Russians (Durand 1900). Durand, the British agent who communicated this offer, was initially treated harshly by the mir, and even threatened with execution. The mir, however, relented when the agent offered to raise the subsidy to 5,000 rupees.

During these talks, the mir repeatedly alluded to reports of a Russian party being dispatched from St. Petersburg, with a message and gifts for him from the tsar. At the time, Durand thought this just empty boasting. But in the lead-up to the invasion of 1891, the mir started sending insolent messages to Durand telling him that "he cared nothing for the womanly English, as he hung upon the skirts of the manly Russians, and had given orders to his followers to bring him the Gilgit Agent's head on a platter"

(Curzon 1926, 187). I suggest that through his use of characteristically vio-
lent and idiosyncratic threats, Mir Safdar was reproducing and deploying
the British representation of him as a savage, barbarian, and lawless ruler,
representation associated with the rhetoric of distance in travel writings.
These figures of speeches were part of the mir's deliberate hyperbolic political discourse.

Europeans had long romanticized the "noble" savage, but this image of
the primitive bespoke fear as well as admiration. The mir's threat to cut off
the British agent's head and send it home on a platter was absolutely the
stuff of Victorian nightmares and imagination. It was horrifying to the
British because of the particularly savage *style* of killing that was threatened. The British were responsible for the killing of many thousands of
people throughout the empire, in Africa and Asia, but their *style* of doing so
was somehow more civilized. For example, Mbembe (2003) describes how
the British in Africa invoked the metaphor of hunting in order to justify
their killing of the natives and in so doing rendered the natives subhuman,
their killing of them somehow civilized.[21] Yet, while the mir's deployment
of these representations catered to British prejudices, to the people of
Hunza they may well have had a different meaning, encouraging local resistance to the British. What is interesting here is that each side used the
same cultural categories for opposing ends: both to justify an invasion of
Hunza and to rally resistance to it.

Meanwhile, the mir played on the deep-seated British paranoia of the
Russians. Evidently he knew, or guessed from the agent's reactions, that this
paranoia was a particularly acute imperial blind spot.[22] Here once again, the
mir sought to turn the cultural phobias of the dominant British to his advantage. To put it another way, we can say that the mir, through this inversionary
strategy, managed to acquire agency within a diplomatic relationship of
asymmetrical power. The co-option of imperial cultural categories by the
mir and the effectiveness of this strategy must, however, be viewed in the
light of both imagined and real events. That is, the British representation of
the people of Hunza as savages and caravan raiders, though based on a mis-
reading of Hunza's struggle with the Kirghiz over the control of grazing
areas, was based on real-life events in which raiders from Hunza did plunder
Kirghiz encampments and occasionally did kill Kirghiz nomads. Likewise,
the mir of Hunza's threat to cut off the British agent's head was no empty
boast given that, since the 1840s, several European explorers and travelers
had met their deaths at the hands of locals in the region.[23]

There was, though, a further way in which the mir interacted with the power of the British: discursively inverting it. During his visit to Hunza in 1889, when Younghusband put the question of raids to the mir of Hunza and asked him to put an end to such practices, the mir refused point-blank. His response, however, left a stereotypical image of him in Younghusband's mind: "There was no diplomatic mincing of matters with Safdar Ali, and his outspokenness did not come from any innate strength of character, but simply because he was entirely ignorant of [his] real position in the universe. He was under the impression that [the] Empress of India, the Czar of Russia, and the Emperor of China were chiefs of neighboring tribes" (1896, 285). At this point, Younghusband asked him if he had ever been to India. This was, of course, a very important question, one that was actually a veiled warning to the mir about the power of the British Empire.[24] Younghusband recalled that the mir, in answer to this question, stated that "'great kings' like himself and Alexander never left their own country!" Younghusband soon became frustrated; he did not know how to deal with a person who seemingly could not or would not recognize the obvious subtext of the discussion, namely, the dominance of the British—which for Younghusband was a naturalized and unquestioned assumption. Younghusband wrote, "The difficulty was, therefore, to know how to deal with such a man as this" (286).

The aim of Younghusband's question was to underline the mir's remoteness and his insignificance to the British; this was a cultural attack. Clearly, the mir could not have denied the reality of his physical remoteness, but he could be more playful and subversive about his "real position" in the geopolitical arena. From the mir's perspective, the whole notion of his remoteness and insignificance as constructed by the British was absurd, given that Hunza was located in the *center* of three empires. Moreover, by claiming his similarity to Alexander, perhaps the most universal war hero, the mir is claiming a universality, just as Younghusband was when he said that the mir was ignorant of his real position in the universe.[25] That may have been true, but the mir was well aware of his strategic geopolitical position in the Great Game.

The British, then, imagined this remote frontier region in cultural and sociological terms as a world whose people, like the mir of Hunza, were beyond reason. The mir, however, did not imagine the British to be beyond reason, or even remote; he spoke geostrategy with them. For Safdar Ali to invoke the figure of Alexander, then, in one of the most memorable encounters of the Great Game, was to create a common language of political discourse because it is improbable that the mir did not know that Alexander did leave his country.

There is no reason to believe that Safdar Ali did not genuinely believe in his own importance geopolitically despite his physical remoteness. In fact, he and others in the region saw their remoteness as a strategic advantage due to the friction of distance. In addition, his resistance to British demands and his attempt to treat the British as equals were perhaps due to the particular history and circumstances in which he found himself. Kashmiri forces, acting at the behest of Calcutta, had tried and failed on three occasions to subdue Hunza. When the British finally managed to establish an agency at Gilgit in 1878, they got cold feet and abandoned it after three years. These failures and vacillations on the part of the British would have given Safdar Ali confidence and inflated his sense of self-importance. The parallel diplomatic courtship of Russian and Chinese officials would have also had an impact on the Hunza ruler's attitude toward British demands. Still, no significant offer of help was promised to Hunza by either the Chinese or the Russians. It is not clear if there was ever a secret understanding between Hunza and China or Russia because local history attributes Hunza's defeat to local defectors *and* unfulfilled promises by Russia. In the end Safdar Ali may well have been simply playing clever but desperate frontier politics against the expanding British Empire.

Conclusion

Beneath the British simplistic and self-serving representation of the Hunza people as a remote society lay a complex web of intersecting geopolitical and local political processes. Despite the difficulties of travel and the social resistance offered by them, the people of Hunza were not isolated from surrounding politics either before or after the arrival of the British. When historical contingency brought them face-to-face with British power, they actively engaged in those politics, negotiating and strategically collaborating or resisting them.

Hunza's remoteness was constructed in the British encounters with and experience of the friction of distance in the region. This experience gave rise to the rhetoric of distance, which was informed by a particular imperial ideology based on the spread of civilization to distant and remote regions of their empire. This rhetoric had a persuasive force that constructed a moral geography of distance: faraway places peopled by dangerous others needed the civilizing interjection of the British.

As the friction of distance was reduced through better transportation systems, Hunza's position shifted from a remote frontier to one of the most important geopolitical hot spots in the region. The British saw the building of roads as the introduction of civilization to the region and thought that local resistance would change to cooperation and coexistence. To some extent that did happen after the British conquest of Hunza: it became a little-known and usually somewhat irrelevant outpost of the British Empire, only periodically important.

After the British conquest of Hunza, Safdar Ali fled to China, where he lived till his death in 1930. He was deposed in absentia and his seven-year-old brother Nazeem Khan was made the mir in the spring of 1892. Mir Nazeem Khan provided the stability the British had long wanted in the region by ruling in an authoritarian manner but remaining loyal to the British—in the process enriching himself and his close allies. The British, in return for the mir's allegiance, turned a blind eye to his internal administrative policies of strictly controlling the population and guarding his own throne against possible challengers.

CHAPTER 3

Frontier Matters

Irrelevance, Romanticism, and Transformation of Hunza Society

With achievement of the dual goals of demarcating the Afghan boundary as a buffer between their empire and Russia's and conquering Hunza, the British had secured access to the main Karakorum watershed. But in the process, they had inserted themselves into local and regional politics. The relationship between the British Empire, the state of Hunza, the princely state of Kashmir, and the Chinese Empire was a complex one that was further tested when the British tried to demarcate the boundary of Hunza with China in the region. Throughout the second half of the nineteenth century, the British consolidated their political position inside the Kashmir state and took hold of its foreign policy.[1] In light of the Russian threat, the British placed a resident in Srinagar in 1884 against the wishes of the Kashmir state. Hunza, which had traditionally been a tributary to China, entered into a tributary relationship with Kashmir in 1870, becoming, by default, a tributary state of the British Empire. The British attempt to settle the boundary of Hunza and China was an example of the hierarchical structure of the British Empire: by fixing the Hunza and Chinese boundary, the British were also determining the boundary of Kashmir and eventually British India.

While Hunza became "peaceful" after being conquered in 1891 and coming under British administration, its official frontier with China would remain unmarked throughout the period of British rule, which ended at partition in 1947. After marking the Durand Line in 1895 on the Afghan frontier and defining the border between the Afghan and Russian frontiers, the central political goal of the British was to nullify Chinese claims over Hunza by pursuing a boundary treaty with China that would bring Hunza under British control permanently. Such a treaty would seal the northern and northeastern borders of the British Empire in the region. Simultaneously, the British worked to wrest administrative and territorial control of Hunza from the maharaja of Kashmir. These two political objectives—to invalidate Chinese and Kashmiri suzerainty over Hunza—however, were marked by ad hocism, vagueness, and lack of full commitment not only on the part of the British but also of the mir of Hunza, the Chinese, and to some extent the Kashmiris.[2]

During British efforts at frontier settlement with China under the reign of Mir Nazeem Khan from 1892 to 1938, Hunza was represented and treated as a peaceful outpost of the British Empire—one that now lacked much geopolitical importance because of a sharp decrease in the Russian threat. The complex, mostly unclear and unsettled status of Hunza's position as a tributary to three competing powers (the Kashmir state, the British government of India, and the Chinese) added to its and irrelevance. The area that twenty years previously had been considered a pivot of Asia, and hence central to the imperial rivalries, had now truly become a backwater of the British Empire. The mir took advantage of the confusions created by the competing objectives and maneuvers of the rival powers. Moreover, the physical remoteness of Hunza added a new dimension of irrelevance and insignificance.

During this period (1890s–1930s) sympathetic British accounts of the people of Hunza replaced the sweeping pejorative descriptions of the past. The new mir was depicted as a benevolent ruler who sincerely believed in his power to call forth rain. One indication of this shift in attitude was that now British political agents periodically asked the local mir to reenact caravan raids. Restaged as cultural performance, caravan raiding was no longer evidence of Hunza's dangerous and threatening primitivism. Rather, it was seen in nostalgic terms as an artifact from a bygone era, a trait of an interesting and rare society, a threatened and endangered other. Hunza's remoteness still served as a metric of difference between the British and the local culture, but appreciation grew among the British for this exotic society. This appreciation was a sign of disenchantment with the excesses of modernization. A glimpse into the lives of British families stationed in neighboring Gilgit town shows the affective dimension of remote areas on the minds of outsiders. This affective quality of remote towns such as Gilgit and Hunza, however, was mediated by the geopolitical context of the times.

Frontier Politics: The Case of Raskam and Taghdumbash

Archival records show that the British political stance on the actual position of the boundary between Hunza and China initially changed in response to perceived or real threats of a Russian invasion into Chinese Turkestan. But as the Russian threat faded away in the first three decades of the twentieth century, there was little urgency on the part of the British to settle the frontier. Nor were the Chinese interested in settling the boundary. Matters were further complicated by the fact that before the British arrival in the region, Hunza had claims to land on the Chinese side of the main Karakorum watershed. The two important tracts of lands in the Chinese territories over which Hunza had usufruct right were Taghdumbash and Raskam. The British used Hunza's claim on these lands as an excuse to extend their influence over the Karakorum watershed. But these claims were weak and lacked political conviction, so little was achieved in terms of boundary settlement between China and the British Empire.

Taghdumbash is an area of high alpine steppe valleys on the eastern edge of the Pamir Mountains beyond the Khunjerab Pass directly north of Hunza in what was then known as Chinese Central Asia or Chinese Turkestan. It is located on the border between Russian and Chinese territories. As mentioned in the last chapter, the mir of Hunza had rights to graze his flock there and collect taxes from Kirghiz and Wakhi pastoralists in this area, which Mir Silim Khan had conquered in the mid-eighteenth century. Hunza's ruler needed the income from grazing taxes collected from Kirghiz nomads. Mir Salim Khan III conquered and brought under his control Kirghiz territory north of the Khunjerab Pass. Mir Silim Khan settled Gojal and gave the land to Wakhi peasants and Kirghiz nomads. This was the same territory that the rulers of Hunza raided to collect taxes and that provided the British with a pretext to invade Hunza. In an ironic twist, the British now indirectly defended these raids when they defended the right of the rulers of Hunza to this territory. In addition to Taghdumbash Pamir, the rulers of Hunza had rights to cultivate land in the Raskam valley east of the Shimshal Pass in Chinese territories. These rights had been recognized by Chinese authorities since the eighteenth century and formed the basis of the political relationship between the two. The mir and the Chinese regularly exchanged gifts, creating a patron–client relationship between the two states.[3] For example, the following was reported in the *Peking Gazette* about the relationship: "Since the year 1878, when the conquest of the New

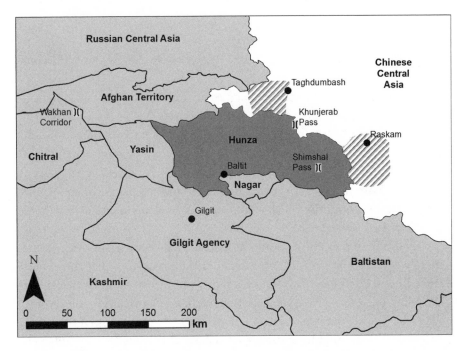

Figure 4: Hunza-controlled territories of Raskam and Taghdumbash (map drawn by Chris Milan)

Dominion was completed by Tso Tsung-t'ang, the Ruler of Kunjut, a small Mahometan State to the south of Lake Sarikol, has been in the habit of paying to China, through the Taotai at Kashghar, a yearly tribute of an ounce and a half of gold dust, in return for which he receives a present of two rolls of satin. The Governor of Turkistan reports that the usual amount of tribute for the present year has been transmitted by him to the Department of Imperial Household."[4]

As early as 1888, the British wrote to the Qing authorities claiming suzerainty over Hunza and asking the Chinese to renounce their claim. It seems that the Chinese Qing were not aware of the political conditions in this remote region. As Lin (2009, 476) writes: "Disturbed by the increasingly close interaction between China and Hunza, the British Minister at Peking addressed the Qing court in the summer of 1888, warning that 'Hunza had long been a feudatory of Kashmir,' and it would be impossible for the British Government of India to allow this petty tribal state to become a tributary to the Qing Empire. The authorities in Peking made no official response, presumably either because Qing policy planners in the court did not genuinely deem Hunza as China's inner dependency, or simply because they were ignorant of what was really happening in the remote Pamir and Altishahr borderlands."

The British backed the mir's claim right after the conquest of Hunza in 1891, even suggesting to him that he should press for his rights in the region. Lin states: "The British sources indicate that Nazeem Khan was advised . . . to resume Hunza's traditional pastoral and grazing privileges in southern Xinjiang. He was even encouraged by policy makers in New Delhi to claim territorial rights over the 'no man's land' of the Altishahr-Pamir border regions, once the collapsing Qing court was no longer able to hold these territories effectively" (2009, 499).

The British took this position in light of their perception that the authority of the Qing rule was crumbling in the region and that this might allow Russia to invade. With Hunza's right (and by extension British right) over the region reasserted, the Russians might think twice before invading. After the British conquered Hunza in 1891, the Russians viewed the occupation of Taghdumbash as a tit-for-tat policy. Ideally, the British preferred Taghdumbash to remain under Chinese control so that the northern frontier of Hunza would not touch the Russian frontier and the threat of Russian invasion would be eliminated (see Chohan 1985). They were, however, also ready to take over the region from the Chinese, who were seen as incapable of defending it against the Russians. Combined with the demarcation of the Afghanistan and Russian boundary in the Pamir in 1895, as seen in the last chapter, and increased British confidence in the Chinese, in the final years of the nineteenth century, the British preferred to give Taghdumbash and Raskam to the Chinese in exchange for Chinese renunciation of claims to Hunza.

In 1899, the British ambassador to China, Sir Claude MacDonald, proposed to Chinese officials that the Karakorum Range should be made the permanent boundary between the two empires.[5] If the boundary was fixed at this watershed, it would mean that Hunza would have to give up ownership or use rights over land on the Chinese side. This official British position was quite contrary to their position in the past, in which they recognized that China had suzerainty over Hunza, as evidenced by the annual tribute that Hunza paid to the Chinese. In the past, the British had actually supplied the gold that Hunza paid as a tribute to the Chinese.[6]

The British simultaneously put pressure on the mir to give up his rights over the Taghdumbash Pamir and the Raskam valleys in return for an annual subsidy. In the event, there was no reply from the Chinese side to this British proposal—at least none is to be found in the records—and the frontier remained undefined.

The mir, likewise, did not pay much attention to this British policy; he continued to press his rights in both Taghdumbash and Raskam. It seems that Mir Nazeem was interested in retaining Taghdumbash and Raskam both as a source of taxes and as a potential bargaining chip with the British. It seems that he was afraid that permanent occupation of Raskam might lead to his subjects of Hunza becoming Chinese subjects. In 1905, the mir asked for compensation for withdrawing from Raskam, and the resident in Kashmir, on the recommendation of the political agent at Gilgit, recommended a compensation amount of 1,000 rupees from the government of India.[7]

The same year, the issue of demarcating the boundary with China was taken up by none other than Curzon himself, by then the viceroy of India, who argued that since China had not replied in 1899, the earlier offer should be regarded as canceled. Curzon made a new proposal of a boundary eight miles north of the Shimshal Pass, in the north of the Karakorum watershed and hence in Chinese territory.[8] In return for Chinese acceptance of these conditions, the British government offered that Hunza would give up claims to the rest of the Raskam valley and all of the Taghdumbash Pamir.[9] Curzon argued that the eight-mile tract beyond the Shimshal Pass should be included in British territory because it was an important grazing land for Wakhi and Shimshali pastoralists who "depend for their grazing almost entirely on the valley between the Shingshal Pass and Dawaza" and it was of "considerable importance to the Shingshalis."[10]

It seems that the Chinese initially agreed to Curzon's proposal; the records show that a representative of the mir of Hunza was told in 1907[11] by the Chinese authorities in Kashgar that the mir would be given an official *san'ad* (deed) recognizing his ownership of some of the side valleys in Raskam valley north of the Shimshal Pass included in the eight-mile area.[12] But the final settlement of the boundary never officially took place, as the Chinese authorities later said that if they allowed Hunza (in effect the British) to take over the land north of the main watershed, then the Russians would make similar demands for Sarikol and the Taghdumbash area in Chinese territory.[13] A status quo was maintained: Hunza herders used Shimshal for grazing but refrained from using the Raskam valleys beyond the Shimshal Pass for cultivation. It is not clear from the records if Mir Nazeem continued to collect grazing tax from the Wakhis and Kirghiz in

Taghdumbash Pamir. The matter appears to have died down until 1912, when it seemed that the Chinese revolution of 1911 would induce Russia to occupy Chinese Turkestan and Sinkiang. The British again revived their interest in Raskam (Lamb 1968, 102), but no major steps were taken or urgency shown to resolve the issue. Two years later, in a letter to the foreign secretary of the government of India, the British resident in Kashmir started by stating, "The correspondence, as will be seen on perusal, relates to the old and vexed question of the Hunza rights in Raskam in regard to which a sleeping and non-committal policy has now been adopted for the last 14 years."[14]

In 1915, the British again took up the issue of delimiting the official boundary with China, and yet again its position differed from its previous proposal. This time the British decided to support Mir Nazeem's claim of ownership not only over parts of the Raskam valley but also over Taghdumbash Pamir, and they did not show opposition to Hunza's effort to forcibly collect taxes from the Kirghiz.[15] Only three decades before, when Mir Safdar had forcibly collected grazing tax from the Kirghiz men in Raskam, the British had called it caravan raiding. Then the British had allied with the Kirghiz; now they seem to be working against them. A year earlier, in 1914, the resident of Kashmir wrote to the government of India: "What now calls for consideration is the fact that the Mir of Hunza, with Major Macpherson's [the political agent of Gilgit] permission, has revived the whole question by dispatching a party of his men to Raskam to evict certain Kirghiz families from the cultivation upon which they have apparently been engaged for the past two years, and take possession of the land themselves."[16]

Hunza continued to be unimportant to the British and the Chinese throughout the first three decades of the twentieth century, as frontier politics cooled down. In the 1920s, the British thought their relationship with the Russians had become cordial, so they did not believe the Russians would raise objections to Hunza's claim on Chinese territory in the Karakorum watershed. They likewise thought that the Chinese were unlikely to show any interest in Taghdumbash and Rasakm as they were "so occupied with affairs nearer home that they will not be likely to display much interest in the somewhat academical question of their shadowy suzerainty over the Mir of Hunza."[17]

But the Chinese continued to claim this shadowy suzerainty over Hunza during this period and threatened to evict Hunza men from Raskam, where they had been cultivating land. In 1922, the Chinese taoyin (previously referred to as "dotai" and "taotai" in official literature) of Kashghar wrote a letter to Mir Nazeem stating that he was "pleased at the action of the said Chief in sending these tokens of his fidelity *from so great a distance* [emphasis added] and compliment him thereupon."[18] Despite this, the Chinese continued to reject any claim by Mir Nazeem over territories on the Chinese side of the Karakorums, perhaps correctly sensing that these claims were actually British claims. In the same year, the taoyin in Kashghar ordered Hunza men to quit the Raskam region, which they refused to do.[19] The British advised Mir Nazeem not to abandon his claim over the region and to adopt a no-response policy.[20]

By the 1930s, the British position on Hunza's claim over Raskam—and, hence, the actual position of the boundary between the British Empire and China—had changed yet again in the light of the anti-British, pro-Russian attitude of the Chinese governor of Sinkiang. As Lin writes, "In the face of a growing anti-British Xinjiang Government dominated by Sheng Shicai, the Mir of Hunza was instructed by the British Government of India not to have any connections with the Sovietized Xinjiang authorities. By 1936, Sheng Shicai's authority had been extended into southern Xinjiang, and his troops, supported by the Soviet Russians, could be seen even as far south as in Tashkurghan. The Mir of Hunza's representative in Sarikol was ousted, and his grazing rights in the Sino-Pamir borders [Taghdumbash and Raskam] were abolished" (2009, 503). This new position (really an old position, in some senses) was similar to the one proposed by Macartney in 1899, in which Hunza was once again asked to relinquish its claim to Taghdumbash Pamir and Raskam.

In 1936, the British tried to forbid Hunza from sending gifts to China. As the political agent in Gilgit wrote in 1937: "The Government sanctioned the proposal in their letter dated 7th December 1936 and imposed the condition that the Mir of Hunza should be advised to refrain from sending that annual present to China in future and to abandon all his rights beyond the presumptive border, such as the right to cultivate lands in the Raskam Valley and to graze cattle and to collect grazing dues in the Taghdumbash Pamir."[21]

The British government proposed that in return for Mir Nazeem's relinquishing his rights over Taghdumbash Pamir and Raskam, his subsidy of 2,000 rupees be increased and he be granted additional land in Gilgit.[22] By 1935 the residents of the Gilgit Agency and, by default, the states associated with it, such as Hunza and Nagar, needed an official passport and a Chinese visa to go across the border. When Mir Nazeem's request to send his representatives was passed on to the Foreign and Political Department, it duly refused to issue the passports or grant permission to the Hunza officials to apply for Chinese visas. Finally, in 1938, the British were able to obtain two goals of their frontier policy in the region, namely, rejection of any claims of Chinese suzerainty over Hunza and an end to Hunza's claim over grazing and taxable land in Chinese territory. That year Hunza paid tribute to the Chinese for the last time.

At this time, access to Hunza and Gilgit remained poor, and not much further geographical knowledge had been produced after the initial thrust during the early days of the Great Game. Even by the mid-1930s, the reputation of the Shimshal Pass had not much changed in terms of its geographical isolation and remoteness. Territories beyond the Shimshal Pass in Raskam, and to some extent in Taghdumbash, remained unsurveyed and unmapped by the British during their rule, and there were "legends of man-eating giants up in Shimshal valley" (Trevelyan 1987, 67). Northern Hunza was considered "one of the most dismal places of the world" and "a spot that was designated 'blank on the map' or 'virtually unexplored'" (65).

The situation from the Chinese perspective was not any better; if anything, it was even worse, and explains why there was no serious effort on the part of the Chinese to settle the issue of border delimitation and demarcation. As Lin writes:

> After the collapse of the Qing, Xinjiang was run as the fiefdom of a series of autocratic rulers only marginally under central Chinese control. From 1912 until the mid-1940s, this vast territory was dominated by the *de facto* independent provincial regimes of Yang Zengxin (1912–28), Jin Shuren (1928–33), and Sheng Shicai (1933–1944). Under the administration of these semi-autonomous governors, only nominal allegiance was extended to the weak central authorities located either in Peking or in Nanking. In other words, given Xinjiang's geographical remoteness and lack of political governability, during the Republican era officials, as well as commoners in China proper, actually knew very little about what was really going on in this territory. (2009, 501)

This practical difficulty of keeping contact with the region was exacerbated by the tactics of Mir Nazeem to extract as much wealth and resources from the British as possible. He often used his claims over Taghdumbash and Raskam to instill diplomatic anxieties in the British. Even in the final year of his rule in 1938, Mir Nazeem reported to the political agent at Gilgit that the Chinese were invading and taking over territory belonging to Hunza up to the Shimshal Pass.[23] According to the letter, the Chinese disputed the claim that Hunza was a British territory. Given the lack of interest and motivation from the Chinese side in asserting any suzerainty over Hunza, it seems likely that Mir Nazeem was inventing this alarming news in order to increase both his own importance and the flow of resources, in the shape of more money to keep border guards at the Shimshal Pass. Mir Nazeem's tactics were in some ways equally influential in dictating the course of frontier politics in the region.

Freeing Hunza from the Kashmir Hold

Throughout the nineteenth century, three powers—Kashmir, China, and Britain—had claimed suzerainty over Hunza, but in the first half of the twentieth century, the British slowly changed this. By 1935, the British had emerged as the paramount power over Hunza, weakening both Chinese and Kashmiri claims over it. In the late nineteenth century, when the British had first contacted the Qing Chinese and challenged their suzerainty rights over Hunza, they did so by stating that Hunza had long been a feudatory to the Kashmir state, which was in turn a vassal state of the British Empire. Throughout their negotiations with the Chinese about settling the northern boundary of Hunza, as we saw in the previous section, the British were also negotiating, both officially and unofficially, the territorial limits of the Kashmir state to the south of Hunza.

Gilgit became a *wizaret*, or a district of the Kashmir state, in the 1860s, about two decades after the formation of the Kashmir state. It was headed by a Kashmiri official called the *wazir-e-wizaret*, who was responsible for the civil administration of the country (Dani 2001, 254–55). Although the British had ostensibly helped to conquer Hunza on behalf of the Kashmir state in 1891, they never considered Hunza to be a part of that state (or the Gilgit Wizaret). Rather, the British considered Hunza part of the Gilgit Agency, which was established first in 1878 and later on permanently in 1889 and was responsible for the external affairs of the Gilgit Wizaret.[24] This was a parallel governance structure to the Gilgit Wizaret. The Gilgit Agency was headed by the political agent and staffed by a doctor and a couple of military officers. The British were transforming the Gilgit Agency from an administrative unit to a territorial one. An assistant political agent was placed in Hunza after 1891 and Hunza was now treated as one of the "political districts" of the Gilgit Agency. In addition to the independent states of Hunza and Nagar, there were regions of Yasin and Ghizer, which were not independent states but ruled by a governor jointly appointed by the government of India and Kashmir. In 1908 Francis Younghusband, then resident in Kashmir, reiterated the position of the British that these political districts were under direct administration of the British and not the Kashmir state.[25] The Kashmir state saw all of these "political districts" as part of the Gilgit Wizaret, its political and administrative structure parallel to the British Gilgit Agency, and hence as part of its territory.

Comanagement of the area continued, but by the 1920s, the British had decided to establish a clear administrative and political distinction between the Gilgit Agency and Gilgit Wizaret in order to limit Kashmir's formal political authority in the frontier region north and west of Gilgit in general and in Hunza in particular. In response to a query from the director of Frontier Circle, Survey of India, regarding how to show the area of the Gilgit Agency on the map, the resident of Kashmir gave the following definition:

1. Kashmir state territory, i.e., Gilgit Wazarat, comprising the Gilgit Tehsil (includes Bunji) and Astor,
2. The Political Districts of Hunza, Nagir, Punial, Yasin Ghizer and Ishkoman and Chilas
3. Un-administered areas of Darel and Tangir.[26]

In the above description of territory under the Gilgit Agency, Gilgit Wizaret was no longer coterminous with the agency; rather, it was just one part of the agency and did not contain the political districts, such as Hunza and Nagar, but did include the area of Gilgit Tehsil, Astor, and Bunji. In the same letter, the resident of Kashmir had also said his office should refrain from showing the political districts of Hunza, Nagar, Ghizer, and Ishkoman as part of Kashmir territory in the Survey of India map.[27] He argued that only Gilgit Wizaret, including the territory of Gilgit district, Bunji, and Astor, should be shown as part of the Kashmir state territory. He also urged the government of India not to use the indigenous term *frontier illaqas* (areas) for Hunza and Nagir but instead to use the British term *political district* of the Gilgit Agency, as the former implied an area not directly administered by the British.[28] The maharaja also continued to protest the British claim that Hunza was not part of the Kashmir territory.[29]

The British finally achieved a semblance of full sovereignty over Hunza when in 1935, right after the India Act, it arm-twisted the Kashmir state to lease the entire Gilgit Agency on a sixty-year lease. It is interesting to note, as argued by Marsh (2009), that the British policy on the northwest and northern frontier region was oblivious to the rise of Indian nationalism; the frontier policy could not anticipate the political end of the empire.

It is not clear how much importance the Kashmir state accorded to Hunza. It seems that the maharaja's claim over Hunza was as shadowy as that of the Chinese. In 1935, when Mir Nazeem was negotiating a subsidy for the loss of grazing rights in Raskam, the issue of his title came up. The British political agent reported that the mir wanted the title of His Highness, but he acknowledged that the maharaja of Kashmir would not acquiesce to this. However, the political agent wrote that while Hunza was not important at all to the maharaja, to the British it was of strategic importance because it occupied a "unique" position. He wrote, "No doubt if his revenues are any criterion he [the mir] is utterly unimportant. About this all I would urge is the unique position of Hunza in the British Empire. . . . If the title of His Highness is still considered out of place I would strongly recommend the grant of KCSI [Knight Commander of the Star of India]."[30]

By the 1940s, the culmination of Indian nationalism, the importance of the Gilgit Agency and by default Hunza had gone off the radar almost completely in colonial state policy and in Kashmir state politics. For the Kashmir state, as for the Chinese, for whom Hunza was a region more in the imagination than in reality, Hunza had little material importance. Kashmir's protests against British political and bureaucratic encroachment on Hunza and the polities surrounding Gilgit Wizaret were more symbolic than real, referring to the British curtailment of Kashmir sovereignty at the heart of the state in Srinagar.

Effects of British Frontier Policy on Hunza

Concern among British officers about the internal stability of Hunza emerged not only from its perceived bad administration by the Kashmir state but also from internal problems in Hunza itself, such as land shortages, deficient agriculture, and low tax rates. This latter fear was aired as early as 1894, when the assistant political agent in Hunza, Lieutenant B. E. M. Gurdon, stated that if the Chinese were given full control of Taghdumbash Pamir, Mir Nazeem would lose his tax revenues from the

herders there and might be inclined to increase taxes within Hunza, which
could lead to hardship for the local people and hence unrest. He stated,
"However desirous the Mir of Hunza might be of increasing taxation, he
would be unable to do so, as, under existing arrangements, the people have
barely sufficient for their daily food."[31]

In the same year, the political agent in Gilgit justified the increase in
subsidy on similar grounds, stating that Hunza (and Nagar) rulers needed
financial help to keep peace in the region by rewarding their followers.[32]

In addition to the practical reasons of promoting peace and stability, the
British support for Mir Nazeem was also a means of projecting their own
power in the frontier region. The political agent continued his letter by asking for the resumption of a subsidy to the mir. "Further it should be remembered that Mohammad Nazeem Khan is a ruler of our choosing, and
for this reason it behooves us to strengthen his position as far as we can."[33]

The abiding concern of the British to maintain stability within the
frontier region of Hunza, which led them to intervene in the internal affairs
of the state, had an important and lasting effect in strengthening the position of Mir Nazeem vis-à-vis the local population. Although the British
expressed concern about local people and were careful to avoid encouraging
excessive behavior on the part of the mir such that the local people might
rise up against him, ultimately it was through the mir that authority and
order were maintained. Mir Nazeem was brought into the domain of local
Indian rulers whose power was both curtailed and enhanced by British colonialism (Ramusack 2004; Hughes 2013). Mir Nazeem Khan's autobiography is full of self-praise, sly digs at the British,[34] and stories about the gifts
he received from British officers in return for his loyalty. He writes about
intermixing with the regional maharajas, the viceroy, and other British and
Kashmiri officials during his "bureaucratic pilgrimage" (Cohn 2004, 63) to
India.

But the British not only brought the mir to the center of power; they
also brought the center of power to the mir through their direct involvement in local politics. The British managed local politics in the Oriental
Darbar (court) style, conjuring the image of "White Mughal."[35] One example is the description of the installation of Mir Ghazan by the British
resident in Kashmir in 1938 after the death of Mir Nazeem Khan. The archival material shows that the British headed the political proceedings in a
native courtlike setting in which they accepted tributes and reciprocated
in the form of subsidies, presented certificates, watched polo games, and

attended local dance parties.[36] They wrote about the Darbar being held on the roof of the Baltit Fort overlooking the terraced fields, with Rakaposhi Mountain standing in the background twenty-five thousand feet tall.[37] The resident in Kashmir described the ceremony: "Later in the morning group photographs were taken in Karimabad (capital of Hunza state), and the government presents were handed out to the Mir, Gushpurs (princes) wazir, etc. Choghas in exchange were at the same time presented by the Mir to myself and other British officers. . . . That evening there was the usual, mounted archery, 'tumbuk' firing and polo, and the Resident and a number of British officers took part in the game."[38]

Most British descriptions of Mir Nazeem Khan, official and unofficial, suggest that he was a caring and kind ruler. For example, Emily Lorimer, the wife of a former British political agent, David Lorimer, writing in 1935, states that Mir Nazeem rendered a wonderful service to the empire by guarding the lofty passes (1939, 121). Lorimer's accounts are somewhat naïve and she does not understand the political economy aspect of the relationship between the British and Hunza. For example, she writes: "When a new work is undertaken for the public good it is supervised by the Mir, or his Wazir acting for him, and it is carried out by the communal labor, of which everyone recognizes the justice and necessity" (83). Lorimer does not mention the Department of Public Works, which was established by the British and headed by Mir Nazeem. He was the sole contractor as well the owner of the land and commanded *kar-e-begar*, corvée labor, from his subjects.

The only person to question the loyalty of the mir—and hence the British policy of supporting him—was Colonel Reginald Charles Schomberg, who traveled to the region on a private excursion but was locally believed to be a British spy. He claimed that Mir Nazeem was sending misleading and dubious reports to the political agent about frontier matters, and that he was misreporting facts in a manner that increased his importance and hence consolidated his position. About the Raskam land and the mir's hankering for it, Schomberg wrote:

> Now whatever rights the Mirs of Hunza once had in present Chinese territory were abolished by the Pamir Commission which delimited the frontier. The British Consular authorities in Kashgar ignore the whole business, and avoid all interference on the subject of the Mir and his rights. It is not easy, however, to abstain from some official recognition

as the Mir of Hunza hardly allows a month to pass without complaining of some petty infringement of his privileges. The whole business is an unmitigated nuisance, especially as the Mir becomes aggrieved and querulous if the British officials do not hasten to support his often excessive demands. (1935, 157–58)

Contrary to Lorimer, Schomberg (1935, 121) stated that Mir Nazeem kept a very close watch on his subjects through a solid spying system and that any instances of dissent or challenge to his authority were quashed with the help of Hunza levies, the local militia, established in 1913 and armed and paid for by the British.[39] Lorimer rejects Schomberg's claim that Mir Nazeem was a cruel and despotic ruler. Citing an example of his benevolent attitudes toward his subjects, Lorimer wrote, "On rare occasions when some mischief-maker threatens serious trouble, the Mir, with the concurrence of his subjects, may deport the would-be disturber of the peace to the distant valleys of Shimshal to tend the royal flocks that pastures . . . food and pasturage are plentiful and life in some respects more luxurious than in lower Hunza. . . . It is amusing to see a recent book indicting the Mir Nazeem for 'tyranny' that imposes so humane and so useful a punishment" (1939, 121–22). She claims that the mir was universally liked throughout his state, which in itself reflected support for British policies in the area.

Evidence from contemporary local accounts and reading between the lines of official accounts suggest that Mir Nazeem was not as benign as Lorimer believed. There was always a delicate balance in Hunza politics maintained by the power of the lineage-based nobility, who claimed a share both of the day-to-day decision making of the state and of its material resources. Moreover, the wazirs mediated the political power between the mirs and the nobility and in fact by the turn of the nineteenth century had become powerful kingmakers in Hunza. They maintained perhaps as much influence over the nobility as the mir and often flexed their muscles to rein in the mir's autocratic tendencies. Mir Nazeem was acutely aware of his reliance on the wazirs for his continued rule. The wazirs were the warrior-administrative class that controlled revenue collection from the peasant class of Hunza. Mir Nazeem and the wazir clashed over chronicling the history of Hunza. Qudrat Ullah Beg, who completed the first locally written history of Hunza in 1935, states that when his grandfather started writing the book (in perhaps the early 1900s), Mir Nazeem commissioned a historian from neighboring Nagar to write a competing history (1967, 3).[40]

The wazirs of Hunza criticized the mirs for their excessive taxation of the local people and for limiting their movement out of Hunza. In 1945, Mir Jamal Khan wrote to Major E. N. Cobb, the political agent at Gilgit, that a full waziri allowance of 600 rupees per annum to Inayatullah should be transferred to his younger brother, Ayesh Khan, and the family allowance of 300 rupees to Inayatullah Beg should be withdrawn altogether.[41] This was a move to weaken the hold of the powerful wazir family of Hunza, who from the time of Mir Safdar Ali had enjoyed relatively more power than the house of the mir himself. Mir Ghazan Khan, Jamal Khan's father and the son of Nazeem Khan, had put Inayatullah Beg in jail along with his three associates. The government of India, however, objected to the full-scale transfer of the allowance of the wazir family to the mir's younger brother on the grounds that the allowance was for the whole wazir family, that is, Humayun Beg and his five sons, of whom Inayatullah Beg was only one. The government of India said that it would be too harsh a punishment for the rest of the brothers of the wazir, who were loyal supporters of the mir.

The introduction of British administration into Hunza made these local power struggles into extralocal affairs over which the local population had less and less influence. After 1891, even the selection of the local mir was taken out of the hands of local clan elders, as the British took over this role. The mir became a symbol of British paramountcy over Hunza, and British rule became a source of power for the mir. The position of the mir had acquired an added dimension of modern institutionalism: he no longer ruled through personal charisma—rather, his powers to invoke rain were now combined with his power to call upon the British-financed Hunza levies in cases of opposition to his writ.

Caravan Raiding and Changing Representations

Before their conquest by the British in 1890s, the people of the northern frontier region were represented in official British accounts as existing in out-of-the-way and dangerous places where civilization was absent. Algernon Durand, the second political agent of the Gilgit Agency, described his work in 1889–90 as "a constant struggle to raise a stretch of frontier 300 miles in length from a condition of incessant war, anarchy, and oppression, into a state of fairly established peace, prosperity, and good government" (1900, vii).

This objective was partly achieved after Hunza was conquered and brought under direct British administration in 1891. In 1892, Francis Younghusband wrote: "I again visited the country [Hunza], to relieve Captain Stewart, the political agent, upon whom had devolved the task of superintending the affairs of the country. At the conclusion of the campaign, the people were quiet and peaceful, as if they had been born and bred under British Administration" (1896, 342).

Once Hunza was brought under colonial administration, a mild sadness, a sensibility of pathos, was evident among the British officers about its loss of freedom. Younghusband writes at this time rather fondly about the intelligence of Hunza men and even expresses some admiration for their practice of "caravan raiding." About Wazir Dadoo, prime minister of Safdar Ali Khan and the main figure behind local resistance to the British in 1891, and his successor Hamayun Beg, Younghusband writes:

> He [Wazir Dadoo] was fond of sport . . . , and of Polo, which is a national game in Hunza. And though the word "raid" was taboo, I could quite imagine him thoroughly enjoying a raid. His successor Humayun, a graver, but equally capable man, certainly did; and when some years afterwards I asked him to show me with a hundred men how he carried out a raid—for he had raided himself—his eyes glistened, and he entered into the proceedings with such zest it was easy to see that the wretched Kirghiz and Turkestani would not stand much of a chance against such a man at the head of such hardy, ready, alert men as these Hunza people were. Later again Humayun spoke to me with a sigh of the "good old days" of the raids. And I had the same feeling for him that I have for hawks. They at any rate have to keep themselves at the highest pitch or they will starve. (1924, 194)

In 1894, three years after Hunza was conquered, Curzon wrote, "No doubt they were great rascals in their free-booter days, but slave-hunting being now out of vogue, they turned the same energies with cheerful alacrity to less questionable pursuits" (1926, 198). One can note the discursive shift taking place in the representation of the people of Hunza in the quoted passages, as the language in which they are described is already less negative. In the same vein, Younghusband wrote this in 1896: "I have felt in looking upon Nature in its aspects wild, in distant unfrequented parts of the earth, and in mixing with strange and little known peoples, who, semi-barbarians though they may be, have often more interesting traits of character than others in a higher scale of civilization" (v). A decade later, Jane Duncan, a British traveler who visited the region in 1905, wrote, "There is a political officer at Hunza which, with its neighbor Nagar, has since 1892 *settled down* peacefully under our rule, after a long career of fighting, robbery and murder" (1906, 3; emphasis added).

In the nineteenth century the people of Hunza were portrayed as a primitive slave-trading and caravan-raiding tribe that *had* to be conquered and brought under the colonial political order (as we saw in the previous chapter). In the first decades of the twentieth century, they had become perfect specimens of the human race, albeit from a Eurocentric viewpoint. Captain L. V. S. Blacker, who traveled to the region in 1920, wrote:

> From the hospitable valley of Gilgit, which, on account of the deep snow and alpine conditions had been reached on foot with a long train of snow-porters, the Mission marched up through the *remote* and rugged valley of Hunza, where it met with a full-blooded hearty welcome in the castle of the Thum. Hunza is inhabited by a fair-skinned people, iron-limbed, Greek in face, Macedonian by descent, endowed with a marvelous physical toughness and an Apollo-like symmetry of form. The tradition of Alexander lives as if he had passed this wild valley but yesterday. Up to the time of their "pacification" in 1892, they had been renowned for wonderful feats of raiding, across some of the greatest glaciers in the world, into the towns of the Chinese. They used to carry a fortnight's rations on their backs and cover immense distances over 20,000-foot mountain ranges, returning laden with gold-dust and jade, driving flocks of Kirghiz prisoners to be sold as slaves. Needless to say, some stalwart specimens of this cheery, smiling-faced people were enlisted into the party. (Blacker 1922, 9; emphasis added)

In 1924, Younghusband described Wazir Dadoo, the wazir of Hunza, as he was in 1889:

> And comparing him with many of the highly educated ministers I afterwards met in the Native States of India—or Indian States, as they are now called—I am not at all sure that Wazir Dadoo was not more capable than them all. There was between him and them all the difference that is between a wild animal and a tame animal. The wild animal

has every hour of his life to depend on its alertness and aliveness: it is in consequence quick and vivacious, and instantly ready to cope with every varying situation as it arises. The tame animal is slow and dull in comparison. Wazir Dadoo had for me the interest of the wild animal. He had all his wits about him. He had the assurance of achieved success. He was in a position of great influence and authority. But he could only retain it with unflagging vigilance. A British Prime Minister leads a fairly precarious existence. He may be in power one day and swept out of it the next. If to these elements of uncertainty there were added the prospect of being pushed off a precipice as well as the loss of power, a British minister would then understand what was Wazir Dadoo's position. (193)

This representation of Wazir Dadoo contrasts sharply with earlier representations in which he and other members of the Hunza elite were shown as unpredictable and irrational, as culturally remote and dangerous characters. By the first quarter of the twentieth century, they had reemerged as quintessential frontier characters, still remote and certainly still fascinating, but no longer dangerous or threatening. The comparison between the British prime minister and Wazir Dadoo in Younghusband's commentary is illuminating here, especially as he was unable to understand the political rhetoric of Mir Safdar Ali. But even now Younghusband's appreciation of the people of Hunza is not based on their *cultural* properties but rather their *natural* properties, which once universalized could be appropriated in British political discourse. For Younghusband, political competition in Hunza was exacerbated by natural competition, which made the lives of local politicians riskier than those of their counterparts in England and therefore made them more sharp and alert. As Hunza became geographically less remote, ironically, it was romanticized and represented as part of nature—and thus in some ways even more remote.

Hunza at this time was also consistently compared with Nagar, a smaller independent state across the river on the eastern side, well inside the empire's borders, to exemplify the superior and non-Oriental qualities of the former.[42] In her book, Lorimer describes the people of Hunza as "non-Oriental people" as "free from inhibiting superstitions as we ourselves" (1939, 82). Notice the contrast between this and an earlier observation of the people of Hunza made about forty years earlier by Lord Curzon: "The superstitious fear of spirits has a far stronger hold than veneration for the

Prophet, and nearly every man carries a number of charms attached to some parts of his dress or person. I even saw them affixed to the leg of a horse. Before the advent of the Europeans these and similar superstitious prevailed to an almost grotesque degree" (1926, 180).

The idea that outside influences through tourism might negatively change remote mountain communities can already be found in popular travel writings in the 1920s, though Hunza was seen as too far in the wilds to be thus threatened. As Jenny Visser-hooft, a Dutch traveler in the area, wrote in 1925, "If there are perhaps regrettable signs that Leh one day may threaten to become a fashionable tourist resort for the jaded globe trotter . . . Hunza is in no such danger. Beyond Gilgit civilization ends" (29).

Only a decade later, however, Visser-hooft's confidence that Hunza would not be affected by modernity and negative outside influences was disproved when Emily Lorimer lamented the fact that cheap Kashmir articles of everyday use had found their way into Hunza and were contaminating its pure and natural way of life (1939, 194). Yet despite the looming threat of outside influence, Hunza was still an unspoiled paradise, according to Lorimer: "My excuse for writing so superficial a book at all must be that it has made me happy to recall this silver honeymoon spent in the Karakoram and to pay tribute to our hospitable and beloved Hunzukuts as they were in 1935, as they must have been for uncounted generations in the past, as they may be for generations yet to come, if the poverty and inaccessibility of their country happily keep them safely quarantined against 'the sick fatigue, the languid doubts,' the unrest and fear and hustle of our civilization" (i).

The movement of products and commodities through improved lines of communication was an indicator of the destruction of distances and the creation of connections. The status of Hunza as a remote area was under threat as it increasingly became connected and transformed, marked by the increased use and display of items and people from elsewhere.

In this new representation of Hunza, its remoteness from centers of civilization is seen as a positive attribute. In it caravan raiding becomes a measure of the physical and mental fitness of the people rather than a sign of their degeneracy. We notice this new trend in the construction and celebration of Hunza's remoteness in the description of travel and routes to the region and in portrayals of the domestic lives of British families in the first four decades of the twentieth century. This trend has continued in its different forms in representations and perceptions of Hunza to the present time.

British Life and Travel in the Region

Toward the end of British rule in India, Gilgit and Hunza had become quite unimportant to the British. In 1943, when Major William Brown was appointed commanding officer of the Gilgit Scouts at the agency, he had never heard of Gilgit before. He tried to gather information from his colleagues who had spent time on the frontier, but "they were all vague. They thought that Gilgit was 'somewhere up in Kashmir' but that was the extent of their knowledge. I determined then that I should go to this mysterious place, this unknown place, a place off the beaten track. Who knew what adventures it might hold. I felt like one of the merchants setting off on the Golden Road to Samarkand" (1998, 23).

Nonofficial records show that the lives of the frontier officers were not quite as romantic and adventurous as Brown had hoped. Yes, there were moments upon which they reflected romantically, but they also talked about daily routine and everyday affairs. Moreover, the "character" of official British was not quite like the way it is upheld in official British discourse. In the British colonial discourse of the time, frontier officers are portrayed as extraordinary men possessing extraordinary qualities of leadership, courage, and political astuteness. In reality, however, these were ordinary people in extraordinary circumstances who had to adjust to the realities of the life in which they were now placed. Personally, they were stingy, groggy, flirtatious, boring, and uncharming as well as kind and friendly. They took notice of local culture as a matter of personal interest as well as routine duty.

We have written accounts from three of the six British families that lived in Gilgit through the 1920s and 1930s. Two have appeared as published books (by Raleigh Trevelyan, son of a British military officer at Gilgit during the late 1920s, and by Emily Lorimer, wife of a political agent in Gilgit in the early 1920s) and one is an unpublished personal diary (Leila Blackwell, family friend of a political agent in Gilgit in the early 1930s). These unofficial accounts reflect the daily routines of life at the agency. Trevelyan's account is a memoir so it reflects his casual and somewhat unguarded thoughts about life in the agency. For example, he wrote: "Contrary to the British legend that the British in remote places abroad even when alone dressed for dinner, Olive and Walter [the Gilgit Agency's military officer and his wife] would dine in dressing-gowns, and I think this was usual on most outposts of the Empire" (1987, 59).

According to Trevelyan, there were a total of six British families, comprising nineteen people (including five children and eight women) residing in the whole of the Gilgit Agency in the late 1920s (1987, 34). Somewhat surprisingly, the women's lives were much less restricted in Gilgit than in many big cities of British India. The women participated in the behind-the-scenes action of the agency, rode horses, went for picnics in the mountains, and had dinner with local rulers. Men played cricket, tennis, and polo and held fishing competitions. They took an interest in nature and observed birds, flowers, and wildlife—and when they shot ibex, markhor, and Marco Polo sheep, they dutifully recorded the length of their heads. Gilgit was seen as an "exceptional" place, despite its remoteness. For example, Trevelyan wrote this about his mother's stay in India: "After Gilgit my parents had gone to Quetta and it was then that Olive had decided that she had enough of India. . . . Olive was above all bored by India, by Indian Army life and, I am afraid, by Indians. She was fed up with the climate and the dirt, unless she could be somewhere exceptional like Gilgit. She adored the scenery, the flowers, the birds, the riding" (1987, 23).

The local rajahs showed the visiting British photos of previous visitors, which reminded them of the special company they were in. They played "God Save the King" on Christmas Day and on the annual jalsas, or gathering of the political heads of the region, held every March. A jalsa was a major event in the agency's administrative calendar, as all the local rajas and mirs of the Gilgit Agency were invited. Tributes and subsidies were exchanged to reaffirm the political hierarchy on a public stage. The jalsa was followed by all sorts of games, including polo and cricket, and music and dance performances. The British romantic view of frontier life was associated with elite and privileged class aesthetics and practices, such as appreciation for the outdoors, observation of nature, and playing sports—and the region provided plenty of opportunity for that.

Conforming to the region's reputation as a remote place that attracted refugees and strangers, during the 1920s and 1930s, there were occasional appearances of White Russian refugees in Gilgit and Hunza. Especially after the Bolshevik Revolution, Trevelyan notes, "lone refugees continued to struggle over the 'high pamere' and Hindu Kush ranges from Soviet Turkestan" (Trevelyan 1987, 54). Some of these refugees were no ordinary White Russians. Trevelyan writes, "In January 1934 there was a flutter in

Gilgit when a party of five arrived, including a man of about thirty who was referred to as 'Romanov' and who claimed to be the Tsarevich, Alexsi Nicolaievich [son of the murdered tsar of Russia]" (55). A companion of his was said to be Prince Sviatopolk-Mirski from Petrograd. He wrote a letter to King George V asking for assistance, which never came forth. His fate remains undocumented in the official records.

The experience of life in the region was especially marked with references to difficult travel conditions and precarious connection and communication with the outside world. Gibson, Luckman, and Willoughbysmith, writing about the remoteness of Darwin, Australia, state, "Remoteness matters for everyone in Darwin at one level, but is also a state of being in which people unevenly invest or believe; everyday life and work is always perceived in relation to other points of reference (wherever they might be)" (2010, 27). Traveling in the Hunza region remained slow and subject to many hurdles. Leila Blackwell, planning a journey to Gilgit, writes: "Anyone with a house to run, had to order all stores in April, so as to have them in Gilgit before the passes were snowed up in October, making it impossible for any large packages to be brought over. . . . Mails are brought over by dak runners, but the men often have to wait in small huts for days, or even weeks, on end for an opportunity to cross the passes" (1950, 2).

The excitement of getting mail is obvious in all the accounts. The event is related in detail, with the condition of parcels and packages vividly described to convey a sense of how far they traveled to reach their destination. Blackwell explains: "People who have never lived in such an out-of-the-way place as Gilgit, often wonder what there is to do. There are no European shops, so that anything required has to be ordered by post from down country, and may take several months to arrive. And when they eventually turn up, the parcels are usually in the most battered condition, no matter how well they may have been packed. In the winter, when the weather is very bad, even letters take a long time to get through, and often the telegraph wires are down for several days on end, thus completely cutting off any communication with the outside world" (1950, 30).

In the mid-1930s Lorimer wrote, "We were amazed to find letters and newspapers coming through to us [in Aliabad, Hunza] in exactly a calendar month from home" (1939, 88). The Gilgit Agency and Hunza were associated with difficulty of access, locations always "behind" other places because modern commodities and consumer items were not available. Accounts of travel to the region are full of suspense, as each journey holds

the possibility of obstruction and the potential for failure. Visitors who succeeded in reaching Gilgit expressed a sense of achievement.

The accounts of difficult traveling conditions, or the friction of distance, were balanced by frequent positive comments about the scenery, even
when it became dreary. Almost all British travelers to the region who passed
through the Burzil Pass described the crossing in some detail. The south
side of Burzil is in a moist, temperate zone marked by high precipitation
due to the monsoon and has conifer forest and alpine vegetation. The north
side is in a dry, high-altitude zone with almost no forest except in the Astor
valley just north of the pass. Trevelyan wrote in 1929: "We were over the
watershed, and there was a dramatic change in the scenery. Everything was
much more austere, barren, but with a new sort of beauty" (1987, 30).
Blackwell, after crossing the Burzil, wrote, "We went through one narrow
gorge, with high cliffs towering on either side—a very grand and impressive
type of scenery" (1950, 10). Lorimer recounts crossing the Burzil Pass:
"After the fertility of the wooded and grassy Kashmir valleys, the bleak
grandeur of the first Gilgit marches, familiar though it was, was finely impressive" (1939, 56). This attitude toward the landscape of the region was
perhaps indicative of how the British saw the region geopolitically. Before
the establishment of the Gilgit Agency in 1889 and the conquest of Hunza
in 1891, most British accounts described the landscape as bleak, barren, and
savage, but three decades later they also thought it quite beautiful.

The perception of remoteness among the British produced a peculiar
response from them. About thirty-five miles south of Gilgit is Bunji, a major garrison town in the 1920s and 1930s that housed both imperial and
Kashmir state forces. Trevelyan recalls Bunji with a sense of nostalgia:
"Bunji was a real oasis for travelers, with green millet fields and poplars, set
in a center of a group of huge mountains. The two British liaison officers
stationed there, Captains Cooper and Eldred, had a pub sign outside their
bungalow, 'Ye Olde Pigge and Whistle,' and there were jokes such as 'Park
your car here' and a signpost showing how many miles to London and
Peking" (1987, 31).[43]

Ye Olde Pigge Whistle was an important symbol of British society in
Bunji. It was an emblem of familiarity, of home, representing a welcoming
instance of British culture to travelers. The presence of a pub, a common
feature in every town in England, in a sea of unfamiliarity and strangeness
evoked a longing for England itself. The great geographical distance (according to the standards of those days) between Hunza and London made

that evocation even more powerful. The absurd milestones at Bunji were meant to be humorous but at the same time signified how far the British were from home, how out of place. Despite their sense of remoteness, the British in the Gilgit Agency lived a carefree and privileged life, and their accounts from this period reflect a sensibility of the pleasure of escaping from modern living conditions. For example, Blackwell wrote, "From all the jollifications that went on, it was hard to believe that we were 400 miles from the nearest railway" (1950, 34). A literal reading of her statement would indicate that happiness and railways exist in close proximity. But the railway here is a metaphor for familiarity and homeliness, and hence what is highlighted is Blackwell's surprise ("hard to believe") at how jolly one can be in an unfamiliar—and remote—place.

Conclusion

During the end of the nineteenth century and the beginning of the twentieth, the British endeavored, eventually with a limited degree of success, to achieve the parallel projects of establishing clear boundaries to their empire and consolidating their authority over Hunza. Although the boundary, in fact, remained unclear, and authority over Hunza remained ambiguous during this period, the region was nonetheless unquestionably "pacified" and brought under British control. The rulers of Hunza, especially Mir Nazeem Khan, used their three-way political relationships with China, Kashmir, and the British as a way of increasing their own wealth and power at the expense of their subjects. The British were perhaps aware of Mir Nazeem Khan's less-than-perfect record of governance, but in the hierarchy of importance, his authority was rated as less corrupt than that of the Kashmiri state—and even as essential to British interests. In the political administration of the frontier region, concern for local and regional governance was always trumped by larger issues of international politics at the imperial level.

Despite their emerging positive attitude toward Hunza, the British remained remarkably ignorant of the region. After four decades of British rule, valleys just outside the vicinity of central Hunza remained "blanks on the map." No major surveys were done in the region by the agency staff, and private diaries and personal accounts written by the spouses of agency officers show that these officers spent their time shooting in the familiar valleys, meeting familiar people, and traveling on busy and familiar routes.

It was only special officers from the Survey and Map Department who visited the most remote places in Hunza. Yet the area of the Taghdumbash and Raskam valleys remained unsurveyed and unstudied despite claims and counterclaims about its ownership. In the 130-year history of the Great Game, not more than a dozen British officers visited the Shimshal Pass.

CHAPTER 4

Rural Romance and Refuge from Civilization

Between the 1940s and the 1970s, multiple treatises were produced on Hunza by British and American medical professionals, farmers, and travelers. In this literature, Hunza emerges as a mythical rural society where people enjoy perfect physical and mental health. The literature argued that the main secret to the longevity and health of the Hunza people was their method of producing food and caring for the soil.[1] Hunza was represented as an ideal rural society, tucked away in its mountain fastness, unchanged and uncorrupted by the influences and values of modern industrial society. Hunza was seen as a haven for rural values— health, simplicity, holism, happiness, sense of community, and self-reliance—that had been lost in contemporary Western rural society under the sway of industrial capitalism and intensive food production.

Commenting on a positive "structure of feeling" associated with the countryside, Raymond Williams argues that "rural virtues . . . in fact mean different things at different times and quite different values are being brought into question" (1973, 12). What can we say about such values in the case of Hunza at this particular point in its history (1940–1970), and why was Hunza's remoteness an important factor in considering these values? Williams states that celebration of and nostalgia for the countryside in Europe represented an "idea of an ordered and happier past set against the disturbance and disorder of the present" (45). Likewise, I argue that literature on health of the people of Hunza is a critique of modernization that points to an apparent contradiction in modern society: the lack of health and happiness despite material progress and plenty. Moreover, I maintain that this contradiction is partially resolved in the presentation of another contradiction: the solution to the fundamental problems created by modernity lies outside modernity; in remote areas we can find the answers to our central problems.

Many of those who wrote about Hunza, however, particularly in the first half of the twentieth century, were inspired less by ideas of antimodernism than by a strictly "scientifically" informed view of the negative effects of certain aspects of changing farming systems and landscapes on human well-being. For these people, too, the isolation of Hunza offered an opportunity to study and understand a seemingly "closed" rural society and its autonomous agricultural system. For doctors in particular, whose challenges to mainstream medical science had to be scientifically valid, Hunza's remoteness and isolation, surrounded as it was by mountain barriers, provided the conditions for rigorous scientific study.

While criticism of modernization and industrial food production may have been valid, the belief that Hunza stood for all the modern world needed was not accurate. The people of Hunza were far from extraordinarily healthy or self-sufficient in food production. The mir of Hunza played a collaborative role in creating and perpetuating the representations of Hunza that we find in the writings of visiting doctors and farmers. We see that the "cultural performances" staged by the mir and his subjects were the key constitutive element in the construction of Hunza as a remote, idyllic rural society. The mir strictly controlled visitors' access to the people of Hunza, ensuring the information they received about diet, exercise, health, disease, and longevity met their preconceptions and expectations. The mir's motive was to keep tight control over his state in the context of increasing outside interference, especially from the state of Pakistan. His power depended on the limitation, or at least control, of that outside interference, and he hoped that foreign visitors would support him in his endeavor to retain that control.

Colonial Health Science

Hunza's status as a place of perfect health first entered Western consciousness through the work of Robert McCarrison, a doctor of the Indian Medical Services (IMS) who worked in the Gilgit Agency from 1904 to 1911 (Wrench 1938, 22), where he had the chance to occasionally treat people from Hunza. About the people of Hunza he would later say: "My own experience provides an example of a race, unsurpassed in perfection of physique and in freedom from disease in general, whose sole food consists to this day of grains, vegetables, and fruits, with a certain amount of milk and butter, and meat only on feast days. I refer to the people of the State of Hunza, situated in the northernmost point of India" (1961, 94).

Doctors and farmers who read McCarrison's work in the 1940s did not visit the area themselves but wrote about Hunza based on his accounts and those of other visitors to the region, mostly explorers and colonial officers. They used primarily anecdotal information as evidence to prove their point about the physical fitness, health, and longevity of the people of Hunza. Later on, during the 1950s, 1960s, and 1970s, increasing numbers of doctors as well as other travelers actually visited Hunza, conducting their research in a "primitivist tourism" (West 2008, 597) mode.

McCarrison had become increasingly interested in the effects of diet on human health during his days as the director of the Nutrition Research Laboratories at Coonoor in the Nilgiri hills (Aykroyd 1960, 415; Arnold 2000, 215). He undertook a series of elaborate experiments on the effect of diet on the health of rats. He believed that the diet of northern India generally, consisting essentially of whole wheat chapatti, milk products, pulses, vegetables and fruit, and some meat, was superior to that of the south, consisting mainly of rice with much less intake of dairy products, protein, or fresh fruits and vegetables (McCarrison 1961, 25).[2] The results of his experiments proved, McCarrison claimed, that the northern diet was much healthier; rats fed on the southern diet contracted a plethora of diseases. Among rats fed the northern diet over a period of five years, McCarrison noted, "there was . . . no case of illness, no death from natural causes, no maternal mortality, no infantile mortality" (26).[3] McCarrison concluded that factors such as ancestry, race, and climate had very little effect on human health, arguing that diet was central; "It may therefore, be taken as a law of life, infringement of which will surely bring its own penalties, that the greatest single factor in the acquisition and maintenance of good health is perfectly constituted food" (21).

McCarrison was out to challenge the dominant paradigm of medical science, which focused on the treatment of disease with drugs, rather than the much more important, in his opinion, focus on health and nutrition.[4] Echoing today's organic farming discourse and also his contemporary in India, one Albert Howard, an economic botanist and director of the Indian

Agricultural Research Station at Pusa who stressed the superiority of indigenous Indian farming methods,[5] McCarrison emphasized the importance of the quality of the soil in which food is grown. "The quality of vegetable food depends on the manner of their cultivation: on conditions of soil, manure, rainfall, irrigation. Thus, we found in India that foodstuffs grown on soil manured with farmyard manure were of higher nutritive quality than those grown on the same soil when manured with chemical manure" (1961, 14).

Healthy Soil, Healthy People

McCarrison's work on diet, agriculture, and health, first published in 1936, proved inspirational for a British doctor, G. T. Wrench. Wrench, while studying medicine in London in the early years of the twentieth century, had been frustrated by the focus of medical practice and research on disease rather than health. He had begun to wonder, "What would happen if we reversed the process and started by learning all we could about the healthiest people and animals who we could discover?" (1938, 6). Recognizing that this was a "ridiculous idea" within the medical profession, Wrench did not pursue it until he read the above-quoted words from McCarrison's book regarding the people of Hunza being "unsurpassed in perfection of physique and in freedom from disease in general." Here, then, were a people upon whom Wrench could focus his ideas about health; his book *The Wheel of Health: The Source of Long Life and Health among the Hunza* was published two years later, in 1938.[6] Commenting on the remoteness of Hunza, Wrench wrote, "There, in a profound cleft, between walls of ten to fifteen thousand feet in height, lies that inhabitable part of Hunza. The beautiful and highly cultivated sunny seven miles, which is the heart of Hunza, may, by its very remoteness, have sheltered primary truths of health which our civilization has *forgotten*" (Wrench 1938, 9; emphasis added).

Although McCarrison had not actually specified that the diet of Hunza was the most healthy he found—his findings from his experiments with rats were based on a more general "northern" diet, specifically that of Sikhs in northern India—Wrench homed in on the people of Hunza, inspired by McCarrison's glowing accounts of their particularly good health. Wrench did not actually visit Hunza himself, relying on McCarrison's work and that of earlier explorers and travelers to the region to substantiate McCarrison's findings.

For Wrench, the health of the people of Hunza was due both to their diet and the agricultural system through which that diet was produced. "We are now able to see with greater clarity and wider observation that the remarkable physique of this people is not causeless or accidental, nor a happy chance of nature, nor due to fresh mountain air, but it has a long history to support it. The inhabitants of Hunza are exceptional agriculturists now, as they must have been in the past, and by their character they have preserved—century by century—a quality of agriculture which has rendered to them through food its return gift of perfect physique and health" (1938, 98). It is clear from the above quote that Hunza is appreciated not for its "natural" condition, as we saw in the last chapter, or its closeness to nature; rather, it is Hunza's culture (history) that is celebrated, a culture that has perfected the art of agriculture over centuries.[7] The object of critique in Wrench's book is the modern food production method under industrial agriculture. Wrench acknowledged that the diet eaten in Hunza was not, on the face of it, very different from the diet of many Europeans; the key difference was the way the food was produced. Wrench quotes McCarrison, who "spoke of the Hunza diet as consisting of 'the unsophisticated foods of Nature'; foods not subjected to artificial processes before they reach the consumer. A 'sophist' is defined in the English Encyclopedic Dictionary as 'a cunning and skillful man, a teacher of arts and sciences for money.' Sophistication for reasons of money does not occur in Hunza" (91).

Under the industrial agricultural method, argued Wrench, food had been "sophisticated," or commercialized, and the entry of money into the equation was the reason for the decline in food quality. Wrench stated that with the rising commercialization of wheat, its appearance, shelf life, and transportability took precedence over its taste and nutritional value; food had become detached from the basic biological functions of the body. This led to a new way of processing wheat in which, Wrench tells us, bran, the skin of the grain, was removed for aesthetic reasons while special milling processes eliminated the wheat germ oil in flour in order to extend its shelf life; the oil made the flour turn sour more quickly (1938, 92). According to Wrench, the sophistication of food had more than a negative effect on physical health; wheat germ oil also provides sexual vitality to the body and calmness to the mind. Thus, according to Wrench, one kind of value—money, associated with modern capitalist society—had replaced another kind—the calmness, vitality, and health associated with "ancient" rural society.

The divorce between the spheres of food production and food consumption was seen by many as the cause of various Western illnesses; during the 1930s constipation was identified as the "disease of disease," or disease of civilization, and was held to be related to changes in preparation and consumption of food under the system of industrial agriculture (Whorton 2000, 1588; Banik and Taylor 1960, 200). Wrench also discusses the way that technological changes had altered the basic structure of soil, using the case of the United States, where increasing population led to the expansion of agriculture and the introduction of steel plows to the fields in 1840. Initially, according to Wrench, the steel plough yielded excellent soil and crops full of nutrition as it dug deeper to unlock the soil organic matter, but as the organic matter started to deplete crops became subject to disease and fluctuating yield.

Another writer inspired by McCarrison, who also read Wrench's work, was J. I. Rodale, an American organic farmer and editor of the journal *Organic Gardening*, which he started in 1942 and which remains one of the most important publications of today's organic movement.[8] In 1947, Rodale founded the Rodale Institute of organic farming in Kutztown, Pennsylvania, which is still going strong. Like Wrench, Rodale never visited Hunza; he wrote his 1948 book *The Healthy Hunzas* based on the literature available to him. Rodale used the example of Hunza to demonstrate his conviction, based in large part on the work of the above-mentioned Albert Howard, that good health depended on a diet of food produced in good soil, fertilized using manure and compost rather than chemicals. "The magnificent health of the Hunza is due to one factor, the way in which his food is raised. Of that there can be no doubt. There may be other, though minor, elements that enter into the situation, but if his soil were different and if his methods of husbandry were not so perfect, this book would have no raison d'etre. The Hunzas might then be something like the Nagyris or the Ishkomanis or the Wakhi [groups living in valleys adjacent to Hunza] and would not be an outstanding example of good health" (44–45).

In Rodale's book we see that McCarrison's generalization about the diet of the people of northern India has now been narrowed down to Hunza, to the extent that even people from neighboring valleys cannot compare with the good health of the people of Hunza. This discursive isolation of Hunza from adjoining societies, as we shall see, is strategic as it gives credibility to the critique that is being launched from there. The remoteness of Hunza becomes an indicator of lack of contact with modernity and thus preservation of authenticity.

As mentioned earlier, this literature constituted a critique of modernity, especially its contradiction that despite progress, modern society fails to meet fundamental human needs such as good health and satisfying social life. Although the writers I have been considering were not radical reactionaries, they considered that the price of civilization had been high. Wrench includes a chapter in his book with the title "Progress by Recoil," arguing that for Westerners to recover the good health enjoyed by the people of Hunza, they must go back in time, a temporal shift represented by the contemporary situation in Hunza. "For progress, therefore, we now have to look backwards. We have raced forwards at too great a speed. We now have to recoil. We have to look back to a period and type of agriculture in which vegetable and animal life were mutually healthy. We have to believe even in the golden age, in which gold did not mean coin in the pocket or blocks in a bank, but an age when the golden sunlight seemed to enter into man through plant and fruit, and bestow the warm gift of health" (1938, 121).

John Tobe, a Canadian farmer, visited Hunza in the late 1950s to study "their way of life" and wrote a book about it. In it Tobe writes: "Often we think of civilization as a synonym for perfection, but even the most civilized man can find things about primitive peoples that are attractive to him—life in the outdoors, habits of dress or social customs. In fact man has found that civilization is a mixed blessing. Many of the greatest problems facing the world today are caused directly by our high state of civilization and the drives it creates. The supreme savagery of a hydrogen bomb, for example, makes the wildest head-hunter look like a small-time juvenile delinquent" (1960, i).

In 1960, an American couple who visited and stayed in Hunza, write: "Time is not measured by clocks or calendars [in Hunza]. Time is judged by the changing of the seasons, and each season brings the feeling of newness, not a fear that time is slipping irrevocably away. In the West, on the other hand, where lives are dominated by clocks and calendars, we tend to view each passing moment as a little piece of life which has cruelly slipped away from us, never to return. Each such slipping bit of time brings us closer to old age and ultimately to death. We worry so much about growing old that we actually increase the process" (Banik and Taylor 1960, 76).

In criticizing Western societies, particularly their agricultural practices, these writers even began to claim that perhaps the roots of "real" society could be found elsewhere, in another age, a remote place and time. The crisis of Western society was the result of a traumatic era that saw the breakup of small farming and agricultural communities in the U.S. Plains during the Dust Bowl period of the 1930s. The crisis was intensely debated within the United States with regard to the values that were being lost, ushering in an era of soul searching. In this process, "distant" and "forgotten places," imagined to be untouched by modern agricultural methods, became repositories of these lost values. The writers imagined Hunza as a place where the destructive power of modern science was unknown and the materialism of the modern age was absent. Earlier, in the first chapter, we noted that Hunza was also represented as the location of the origins of civilization and the abode of an original Aryan culture. In this antimodern critique, it was now seen as the original home of agriculture.

Hunza's representation as an ideal rural society was a continuation of a long-established theme of tension and nostalgia relating to rural versus urban settings in Western society within the framework of changing social and economic conditions in the late nineteenth and early twentieth centuries. This continued industrialization of agriculture, with its "high-modernist" (Scott 1998) focus on standardization, mechanization, and commercialization, and the concomitant decline of the small-farm tradition, was blamed by many for the 1930s "Dust Bowl" disaster and the wider Depression (Worster 1979). Steinbeck's *The Grapes of Wrath* and McWilliams's *Factories in the Field*, which both came out in 1939 and became nationally known, summed up the situation in the grimmest terms. This was the backdrop of experience against which Hunza was constructed as an ideal rural society.

Mountains, Isolation, and Natural Experiment

Mountains hold particular significance for European romantic sensibilities about nature. Once seen as fearsome locations of danger and savagery, during the eighteenth century, mountains began to undergo a change in their representation, coming to be viewed as sublime, their vast, wild vistas associated with God's power and truth (Nash 1967, 45–46; McFarlane 2003, 15–16). A more pragmatic, though related, view of mountains emerged in India under British colonialism with the establishment of numerous hill stations during the nineteenth century. Hill stations provided places where the British could escape the heat of the plains during the summer months and enjoy the salubrious climate (Moore, Kosek, and Pandian 2003, 22). This significance of mountains is evident in the literature on Hunza that we have reviewed thus far, but it is not the mountains themselves that are important. Rather, I suggest that the particular power of Hunza's mountainous location in the idyllic rural society discourse has to do with isolation. That is, Hunza's isolation from the world, due to its being "locked" behind mountain barriers, is what gives it its power in the discourse I am discussing.

This isolated nature of Hunza is mentioned far more often than its mountainous environment itself. In the context of criticism of Western industrial agricultural practices, Hunza appeared as a society that, by virtue of its mountain barriers, had remained untouched and thus uncorrupted by modernization. The writings tend to present Hunza as Shangri-la, the mystical and perfect (and imagined) valley populated by a people of unheard-of longevity described in James Hilton's 1933 novel *Lost Horizon* (which numerous travel companies selling travel packages to Hunza today claim was inspired by Hunza). The idea of an isolated paradise barricaded from the outside world also conjures up notions of the Garden of Eden; Grove (1996) writes of the relevance of a search for Eden in the tropics to colonial practices. He argues that northern India became a favorite candidate for the location of the Garden of Eden after the original focus on islands, contending that the penchant for building botanical gardens in the colonial period reflected this Edenic search (39). Using the language and metaphors of earlier accounts from colonial-era geographical exploration and discovery, mountains, thus, became the barrier that kept Hunza undiscovered. As Wrench writes, "Everything suggests that in its [Hunza's] remoteness it may preserve from the distant past, things that the modern world has *forgotten* and does not any longer understand, and amongst those things are perfect physique and health" (1938, 20; emphasis added).

The region's remoteness and the impenetrability of the mountain barrier as reasons for an unchanging situation in Hunza are recurring and persistent themes in the literature. The older discourse of friction of distance, represented by inaccessibility and terrible road conditions, is resignified in the (new) discourse. For example, in 1955, Ian Stephens, an Englishman and a gentleman traveler, wrote, "It is the very terror of this road that has allowed Hunza to exist in serene isolation, safe from more powerful neighbors and untouched by modern civilization" (173). Hoffman also used the discourse of friction of distance to explain Hunza's lack of change in agricultural practices. "Since Hunza has been more or less isolated from the rest of the world, the methods of planting, growing, and harvesting crops have not progressed. One of the main reasons for this is that it is impossible to take into Hunza any of the modern implements used in the western world" (1968, 37–38).

Change and progress, which had earlier been synonymous with movement toward civilization, now became indicators of disorder and disturbance. Preservation and tradition became the new ideals by which a society was to be judged; ironically, the well-being of a society was evaluated on how little it had changed, not how much. In the 1870s, John Biddulph, the first political agent of the Gilgit Agency, wrote that the people of Hunza had not changed in fourteen hundred years and had thus remained dangerous savages. In 1958, John Tobe, the Canadian farmer who visited Hunza, wrote: "Hunza is one of the few places in the entire world, where there is a people who have retained their character, independence, freedom and primitive way of life. It is my belief that they are not much further advanced today than they were more than 2,000 years ago . . . they owe their strength, their health, their longevity, their happiness to that wall of mountains that surrounds them" (1960, 266). Banik and Taylor wrote in much the same vein: "The few arable acres of this tiny country [Hunza] have sustained its meagre population for nearly 2,000 years in almost complete isolation from the rest of the world. Whatever their ancestry, the men of Hunza have been able to maintain their isolation and today live in health and happiness to the age of 120 years" (1960, 24–25). They go on, "The fact has been verified that the Hunza civilization dates back more than 2,000 years and those conditions now are relatively the same as they were then" (26).

Hunza's isolation from the rest of the world is not only relevant insofar as the valley appears as a mythical Shangri-la, however. In most of the writings I have discussed, there is a strongly *scientific* bent to the arguments made. These were not simply travelers seeking to find a perfect society; they were doctors and farmers seeking to demonstrate the validity of their challenges to the wider medical and agricultural scientific community. Like Curzon and Wood, these were also eminently rational people, so it would be a mistake to draw too clear a line between their idealistic sensibilities and a purely rational, scientific outlook. Many scientists, particularly doctors, in the late nineteenth century easily combined a dedication to scientific approaches with concerns for nature and society, both of which they saw as threatened by some elements of the process of modernization. A group of Scottish doctors working for the East India Company in the nineteenth century, for example, was a key advocate of environmental protection in India, arguing that the degradation of the environment had negative impacts on human health. They were inspired, argues Richard Grove, in part by Alexander Von Humboldt, whose ideas about the link between human beings and nature were, in turn, based on holistic Hindu notions of the place of humankind in the natural world (1996, 11). These doctors also made a connection between their environmental concerns and wider morality, as Grove writes: "Underlying their environmental critique, a radical and reformist message was being articulated in which a real concern about environmental health was used as a vehicle to express anxieties about the social and moral consequences of colonialism and industrialism" (462).

These doctors worked for the same IMS that McCarrison worked for in the early twentieth century; thus, their ideas were not necessarily as radical a break as it might appear. In the general drift of their criticism, aimed at the wider systems of modern industrial agriculture and capitalism, McCarrison, Wrench, and others were attempting to make a specific and far-reaching case against the scientific paradigm on which Western medical and agricultural sciences rested. Thus, while they set out to challenge it, at the same time, they had to present methodologies and findings that would be taken seriously within that paradigm. In this context, Hunza's isolation rendered it a natural laboratory in which the effects of a non-modern agricultural system and its effects on the human body could be studied scientifically.[9] Rodale, for example, presented Hunza as a natural experiment from which the scientific community could discover beneficial agricultural practices and secrets of health. "Nature has provided us with an Experiment Station ready-made and with the results that cannot be neglected. Perhaps in the years to come, some heaven-sent investigator of the Charles Darwin type will go thoroughly into this Hunza question on the spot, and will set out clearly all the factors on which their agriculture and their marvelous health depends" (1948, 81).

At around the same time Hunza was being extolled in this literature, the mountain communities of Abkhazia in Georgia and Cochabamba in Ecuador also became popular for study because their people enjoyed incredible health, as Wrench and Rodale noted. In all these cases, isolation and barriers to entry discursively constructed these societies as living laboratories in which a particular theory of nutritional cycles could be demonstrated. Terrie Romano (2002, 2) in her study of Sir John Scott Burdon Sanderson, father of modern pathology and clinical psychology, states that by the end of the nineteenth century, laboratory experimentation became the mainstay of medical education in England. Laboratory-based research emerged as a separate field of biological study distinct from naturalist research, which was still founded on "studying nature" in the nineteenth century.

In this chapter I show that these two methods come together as nature becomes an actual laboratory. Western medical science since the Victorian era had developed a strong convention of empirical observation based on experimentation in a laboratory setting (Arnold 2000; Grove 1996; Romano 2002, 176); the longer the period of experimentation, the stronger the claim to authenticity and validity. Prior to this, science had been seen as the pursuit of gentleman scientists who carried out observation and empirical work in natural settings—hence the rise of the natural history sciences. By the end of the nineteenth century, as science became more laboratory based, medical science took the lead in this regard. The insistence on the idea that Hunza had not changed since time immemorial due to mountain barriers was necessary to construct it as a long-term scientific experiment in a natural setting. Hunza society was constructed as a control group necessary to make comparisons with the effects of modernization on human health in Western societies.

Sources of Evidence

Despite the desire by doctors and farmers to provide scientific evidence of their theories about agricultural methods and health, the vast majority of the data provided is not only anecdotal and at best circumstantial but also often questionable with regard to sources. Both Wrench and Rodale use only secondary and anecdotal evidence taken from the writings of earlier explorers and travelers. For example, to back up his claim about how physically fit the Hunza people are, Wrench quotes Aurel Stein, a British artifact hunter who passed through Hunza in 1901, regarding the remarkable stamina of a Hunza native: "The Messenger had started on the 18th. It was just seven complete days between his start and his return, and in that time he had travelled two hundred and eighty miles on foot, speeding along a track mostly two to four feet wide . . . the messenger was quite fresh and undisturbed, and did not consider that what he had done was unusual" (1938, 11). Rodale recounts a story by R. Skrine, who visited Hunza in 1922. "So it was that Mr. Skrine saw the Mir of Hunza at polo when nearly seventy. As Captain of his side, after a goal, he had to gallop at full speed half-way up the ground, fling the ball into the air and smite it towards the opposing goal" (1948, 108). Rodale also quotes General Bruce of the Royal

Mountain Battery, who describes the people of Hunza as the "best slab and rock climbers in the world" (19).

Those who went to Hunza after the 1950s used similarly anecdotal "evidence" to prove the extraordinary health, physical fitness, and longevity of the people of Hunza. Banik and Taylor watched a volleyball game between younger and older men of the valley, the younger aged between 15 and 50 and the older all apparently over 70 years old, one of them as old as 125. "Both teams played a strenuous game in the scorching heat of the afternoon sun. If any player was fatigued at any time during the game, it was not discernible. They all seemed as relaxed and comfortable as though they were playing a friendly game of canasta" (1960, 45). Jean Shor wrote: "Longevity is a national characteristic. While the *arbab* of Misgar was in his sixties, some of his advisers were pressing a hundred. Many of the diseases of civilization, including cancer, are unknown. It seemed as if hardly anyone dies in Hunza unless he falls off the incredibly narrow trail that links the village down the valley and is dashed to death on the rocks below" (1955, 267).[10] Although his entire book is dedicated to learning how to live to be as old as the people of Hunza, the only evidence Hoffman provides of their age is the following: "During our conversations with the older people, I always asked them how old they were, and many of them said that they were more than a hundred years old" (1968, 49). Dr. Alexander Leaf, a Harvard Medical School professor, recounted an incident when he was shamed by not being able to keep up with a 106-year-old local man on a six-hour mountainous hike (1973, 111).

The reliability of the "data" on the central points of the "research," that is, age and disease, remains problematic. First, there were no birth records in Hunza at the time these reports were produced. Unlike in other areas of India, the British had not put in place a bureaucratic setup to collect and maintain official statistics. Hunza remained an autonomous state until 1974 so the Pakistan administrative system had not yet been introduced. The extraordinary old ages of people were generally accepted as reported by the people themselves or by guides and hosts, with only a few efforts made to verify the truth of these assertions. For example, Stephens writes: "I was introduced to a vigorous grey beard, brisk of step and firm of handclasp, who, judging from his replies to questions about events in history of Kashmir State, must have been 97—as was claimed. He looked no more than 65" (1955, 170).

Claims of the remarkable health and longevity of the people of Hunza were actually refuted by a survey conducted in 1955 by a team of Japanese doctors who argued that the belief in the absence of disease in Hunza was incorrect. The survey report stated: "In reality, the Hunza Area does not appear to be a Land of God's Grace, and it is not different from any other land in the world. Harada found many people in the area suffering from tuberculosis, conjunctivitis, rheumatism, goiter, skin disease, etc. He noted the high rate of infant mortality which was the result of undernourishment" (Imanshi 1963, i).[11]

Interestingly, this report seems to have had absolutely no impact on the continued projection of Hunza as the land of the healthiest people in the world. Why did these writers accept so uncritically the extraordinary ages of the people of Hunza, most of which were in fact provided by the mir? Most researchers did not address the issue of the veracity of their claims or the methodology by which they adduced their "evidence." Those who did were unconvincing.[12]

These writers went with certain preconceptions of what they would find in Hunza, based on their ideological positions with regard to mainstream medical and agricultural practice, and the conclusions they arrived at were heavily shaped by these beliefs. Another reason the visitors seem to have so readily believed the fantastical claims about Hunza's health lies in the agency of the local people, who sustained and coproduced these ideas about the superiority of Hunza's agriculture, the health of its people, and the general well-being of its society. The ruling elite of Hunza, mainly the mir himself, mindful of their representations in the past as constructed by colonial powers, now strategically represented their society as they wished it to be seen (see also Turner 1995, 105).

Sociopolitical Context of the Mir's Rule

During the first half of the twentieth century, the colonial government granted land and money to the rulers of Hunza for building new irrigation channels. Most of these lands were in the Matam Das and Oshkhan Das hamlets of the Gilgit Agency and were given as compensation for land relinquished to Chinese Turkestan. During the building of irrigation canals, the mir's subjects converted hitherto undeveloped *das*,[13] or meadowland, into flourishing agrarian landscapes. These new settlements meant increased revenue for the mir. Labor remained the critical factor in this expansive phase, and it was during this period that the rulers of Hunza devised strict policies to control the internal supply of labor. By the 1930s, check posts had been erected at the southern boundary of Hunza, where Hunza levies controlled the movement of people across the "border." Oral history records show that only in very rare cases were people allowed to leave Hunza; the mir argued that their interaction with outsiders would expose them to dangerous and seditious ideas. In 1935, Mir Nazeem Khan even refused to let his wazir's son go to Srinagar for an education because of these reasons (Beg 1967, 2).

The mir's policy of keeping a close eye on his subjects was a result of the frontier politics in which the rulers of the state of Hunza found themselves. The rulers of Hunza have traditionally played the card of their strategic location—as the guardian of the frontier—very well, thus drawing much-needed resources and importance to themselves. They used this position also to leverage whatever sovereignty and independence they had left after 1891 so as to at least appear to their subjects to still be rulers. (This was a predicament experienced by virtually all Indian princes during the British Raj.) For Mir Nazeem Khan, keeping tight control over information about the frontier condition became a long-term policy of his forty-six years of rule. Schomberg relates one tactic the mir used when the British tried to open a public post office in Hunza in the early 1930s: "So the Mir arranged for the general boycott [of the post office] and, after six months during which not a single letter was posted nor a single postal transaction had taken place, the office was closed. At present, the Mir himself receives the Dak or letter bag, and takes a paternal interest in all correspondence" (1935, 116).

Despite Mir Nazeem Khan's efforts, his policy of control of his population during the 1930s was not absolute. Some were able to leave, particularly in 1935 when the British acquired the Gilgit Agency (and by default Hunza) on a sixty years' lease from Kashmir, and a number of avenues of out-migration were opened to the people of Hunza. Before 1935, it was illegal for someone from Hunza—except the mir, of course—to purchase land outside Hunza. After 1935, the British allowed this. Some people from Hunza found jobs in newly opened employment sectors such as the Gilgit Scouts, the hospital, the Public Works Department, and the Education Department. Others, particularly members of the house of the wazir, started to make contacts with the courts of their spiritual leader, the Aga Khan, in Bombay (Beg 1967). Overall, however, the mir remained in control; for example, despite the efforts of the Aga Khan, the mir did not allow the house of the wazir to build a *Jamat Khana*, the Ismaili Muslim house of worship, in the capital, Karimabad, and his control was generally supported by the British. For example, Beg recalls an incident in 1937 when people from lower Hunza (non-Burosho) brought a complaint to the political agent against the mir; the agent handed over the people to the mir, who threw them in prison (29). It is interesting to note that, when the wazir of Hunza requested the Aga Khan to intervene on this issue and order the mir to allow a Jamat Khana to be built, the Aga Khan advised him to remain patient and follow the wishes of the ruler of the country (34).

In 1948, Hunza, along with other districts under the Gilgit Agency, fought a war of independence against the Kashmiri (and Indian) forces, and Mir Jamal Khan[14] opted to join the newly formed state (1947) of Pakistan, albeit still as a semi-autonomous state, governed as part of the Gilgit Agency. Although the Pakistani state in general supported the rule of the mir, he remained a potential threat to its power. During the 1950s and 1960s, Hunza was further integrated into the Pakistani economy and politics, and many of its inhabitants sought economic opportunities and an escape from the rule of the mir in Hunza in cities like Karachi and Lahore. Here these people engaged with the labor movements and other social movements formed by minority groups in Pakistan. Many people of Hunza joined national political parties and started to demand the abolishment of the mir-dom in Hunza and full integration of the state into the federal structure of Pakistan. These threats to his power added to the insecurity of the mir, who had already begun to lose power after Ismaili religious institutions had finally been allowed into Hunza in 1944.

During the period under review, visits to Hunza by foreigners were strictly controlled by Pakistani authorities. Not only were foreigners under suspicion of being spies, the Pakistani authorities also did not entirely trust the mir. Although on the surface relations were smooth, official intelligence reports show that deep mistrust still persisted in the capital city Karachi (later Islamabad) about the loyalties of Hunza rulers. As a result of this general atmosphere of mistrust, very few foreigners were allowed into Hunza. After the 1950s in particular, when Hunza came under the direct administration of Pakistan, special permission was needed to go there, and it was notoriously difficult to acquire the necessary papers to enter Hunza between the 1950s and the early 1970s. Those lucky few who did manage to get a permit were able to do so only through connections in high places. Hoffman, for example, had to go directly to the president of Pakistan. He wrote: "It is almost as hard to make a trip to Hunza land as it is to take a journey to the moon. While thousands of people have tried to get permission to go, except for a few, the vast majority have not been permitted to enter the country. The reason why it is so difficult to enter is that it borders communist China. The king, or Mir, and the Pakistani government with which Hunza is allied, do not want to provoke the communist countries north of them by permitting westerners to enter Hunza. . . . In order for one to get permission to go to Hunza land, he must first receive an invitation from Mir Mohammad Jamal Khan. He must then write to the Pakistan Minister of Kashmir Affairs in Rawalpindi, requesting permission to make the trip. However, . . . the chances of obtaining this permission are quite remote" (1968, 5).

Here, clearly, the remoteness of Hunza is constructed in the political process. These administrative barriers enhanced its inaccessibility to out-siders and added to its mystique. Oral records based on interviews with elderly people in Hunza show that the mir was ordered by the Pakistani government to keep a strict eye on foreigners who were given permission to visit Hunza and to report their activities to the Pakistani authorities. The mir saw an opportunity in this state of affairs to try to improve his own situation. He also managed to procure a small but not insignificant source of income from the visitors; as the ultimate owner of labor in Hunza, he received a fair amount of cash payments from services his subjects rendered to these outsiders. This was particularly true of visitors who came as mountaineers, for whom considerable porterage was required.

He carefully "managed"[15] the outsiders, not only keeping a close eye on them to satisfy the Pakistani authorities but also making use of his control over them to bolster his own weakening position of power within his king-dom. While the mir's control of the internal population of Hunza was in-creasingly fraught with anxieties and ineffectiveness, and he was faced with the imminent danger of Hunza's further absorption into Pakistan, the few visitors were one group over which he could exercise almost absolute control, while they were in Hunza, at least. The mir made use of this by exploiting the visitors' symbolic value as representatives of external powers to demonstrate his own power within his realm, and also cleverly manipu-lating the visitors' impressions of Hunza in such a way as to strategically present an image of Hunza to the outside world that supported the position of the mir as the all-powerful leader. At the same time, the mir had to be careful that the visitors did not expose his subjects to ideas that might en-courage insubordination; while he wanted visitors to come, he did not want so many to arrive that he could not control them. He was supportive of the restrictions on visitors to the valley. He wrote in the foreword to Hoffman's book, after welcoming Hoffman's endeavor to find out the "truth" about the longevity of his people, "Unfortunately we cannot grant entry to the hundreds of people who want to visit our country. Due to political reasons, we can only admit those who have very urgent and valid reasons" (1968: ix).

In order both to limit the extent to which visitors could share their ideas with the people of Hunza and to present to the world his desired image of Hunza as a remote, ideal rural society—which required that the visitors did not receive contradictory impressions—the mir took care to ensure that communication between the visitors and his subjects was restricted. Visitors were put under the direct supervision of the mir and given very little chance to interact with the general population. This policy began with the first Westerner who lived in Hunza for research, ex–political agent Colonel D. L. Lorimer, who was assigned a member of the house of the wazir, Qudrat Ullah Beg, as his official guide. In 1953, Jean Shor was assigned the governor of Misgar; in 1958, John Tobe was assigned the brother of the mir; in 1961, Jay Hoffman was assigned the uncle of the mir and so were Banik and Taylor. Western outsiders were usually assigned to a member of the ruling family or of the house of the wazir, who acted as their permanent guide during their stay in the country. All foreigners stayed in the guest quarters of the mir's palace as his personal guests, and their most important informant was the mir himself. As we have seen, Mir Jamal Khan even wrote the foreword to one visitor's book, Hoffman's, while in several others, he is often the only person directly quoted. The physical movements of the outsiders were controlled; under no circumstances were they allowed to wander off and meet people freely. One can imagine that these foreigners who did not speak a single word of Buroshiski were totally at the mercy of state officials, who were thus able to present whatever picture of Hunza they wanted to produce.

The mir played a central role in perpetuating both the idea that his people lived to extraordinary old ages and that Hunza represented an authentic rural culture. In the foreword to Hoffman's book, in which he praises Hoffman's efforts to understand the reasons behind the remarkable health and longevity of his people, the mir claims, "Counted by the same calendar used in the Western world, many Hunzakuts have lived to be well over the century mark: from 100 to 120 years, and in isolated cases to as much as 140 years" (1968, viii).

During their visit, amazed at the ability of the older men, Banik and Taylor once asked the mir how the old people were able to remain so physically fit and mentally agile. The mir replied, "When will you people learn that our men of 100 feel no more fatigued than our men of 20? Be careful what you say, or soon you will have our people of over 100 feeling three times their age. And then they will think they are growing old" (1960, 45). The mir seemed to allude also to the beneficial political effects of living a simple and carefree life, because it does not attract the interest of powerful players and so allows a people to keep their independence from outside intervention. "We are the happiest people in the world. We have just enough of everything but not enough to make anyone else want to take it away. You might call this the happy land of just enough" (46).

In both these comments to Banik and Taylor, the mir's underlying message seems to be that his people (including himself as ruler) and Hunza generally should be left well alone, conceptually remote from outside corrupting influences such as greed—and too much self-awareness. The mir suggests that part of the reason for the "special" nature of the people in Hunza is that they are not themselves aware of how special they are. Thus, to remain special, they need to remain ignorant of the outside world, but this ignorance is translated here into simplicity. But also at stake is a threat to the mir's own rule and position. He presents his interests as synonymous with those of the people and charms his visitors with images of rural innocence and bliss. Writing about community development in contemporary Indonesia, Tsing states that in their efforts to draw resources from development organizations and the state, communities do more than just deploy an image that they think is strategically appropriate; they also create a seductive field around their image. These fields of seduction are in effect "fields of power" that are created and controlled by communities to keep the balance of power for "collaboration" in their favor (1999, 162). Likewise, the mir is playing his cards carefully, maintaining the interest of the researchers, which depends on their seeing a particular representation of him and his people, but at the same time aware that this representation could collapse if the researchers find out too much or are exposed to the "wrong" information. In his study of British information systems in colonial India, Bayly shows that the British faced considerable problems in obtaining correct information from frontier regions, impeded not only by the mountainous terrain but by "the wary and tightly knit Nepali elite [which] was . . . able to control the flow of information" (1999, 100). So although visitors to the remote Hunza valley were actually physically in the area, the mir's efforts kept them in many ways distant from it. Despite being among the people of Hunza, it was almost impossible for outside visitors to get close to what the real Hunza was like.

Conclusion

In this chapter I have argued that the construction of Hunza as a remote, rural, and idyllic society was imagined in the critique of Western modernization and civilization,[16] especially the West's basic contradiction: that it fails to satisfy elemental human needs. The chapter shows that the people who constructed Hunza as the ideal rural society were not simply romantics; they were often scientists who used romantic language and discourse to critique industrial agriculture and, indeed, the wider capitalist ethos that had become, in their view, the defining characteristic of Western society.

What was being critiqued was not only the effects on society of the Western system of the production and consumption of food, but also the ubiquity, or inescapability, of this system. The emergence of industrial agriculture in America and Europe in the first half of the twentieth century was remarkable in its pace and scale. The expansion of industrial and commercial agriculture and the decline of small farms, the stronghold of American rural identity, had seemingly sounded the death knell of a viable rural society. Indeed, it was the literal as well as the metaphorical remoteness of Hunza that was exploited and brought into play by critics of industrial agriculture. Hunza was not just any rural society—rather, it was a remote rural society, an original conceptualization, full of moral and persuasive force. Hunza's remote location made its rural status that much more appealing to those who came looking for an uncorrupted and innocent society.

Half a century before, the isolation of Hunza due to its mountain barriers was seen as the reason for the savagery of its people, the cause of violence and resistance to progress. In the late nineteenth century, the British deliberately described Hunza as a community of caravan raiders. Remoteness led to stagnancy, or even degeneracy, and it was only disturbance from the outside that would enable progress to take place within Hunza society. Fifty years later, the same mountain barriers were seen as protecting Hunza from the curses of modern civilization. Rather than Hunza's state of nature threatening outsiders' state of civilization, it was outsiders who became a threat to Hunza's perfect rural society. Remoteness had become a virtue; disturbance had come to be seen as something threatening.

CHAPTER 5

The Origin of a Nation
Hunza and Postcolonial Identity

In the summer of 2005, I was sitting on the sidewalk of a steeply climbing street, made of ancient granite stone, leading up to the foot of Baltit Fort, the eight-hundred-year-old residence of the mir of Hunza. I was talking to an elderly Burosho man, trying to convince him that the British had deliberately misconstrued the Hunza state's tax-collecting practices as a habitual criminal activity of caravan raiding. The Burosho man disagreed with me, insisting that such representations were correct; that the people of Hunza had indeed been criminals and caravan raiders in the past and that there had in fact been a special "department" within the Hunza state responsible for organizing these raids. According to my informant, the department was called Dewan Baigee, a Persian phrase that translates as "the Collector of Taxes." Since the name of this "department" reflected my interpretation and historical explanation of caravan raiding, as discussed in chapter 2, I tried to explain the logic behind it: yes, indeed it was a department overseeing tax collection, sometimes by force, that is, raids. My companion was unimpressed by my explanation and uninterested in my problematization of the people of Hunza's past representation by the British. His was an example of the almost universal response I received when I tried to reinterpret Hunza's caravan raiding past to local people. The same representation is made by local historians who, it seems, rely primarily on colonial sources (Esar 2001; Hunzai 1998).

What was I to make of this insistence by the people of Hunza on unproblematically embracing their historical politically created negative

representation as fact? Why don't the people of Hunza denounce such negative accounts and adopt the truth of their past? I argue that the persistence
of the myth is part of a wider public discourse in Hunza that highlights
their improvement from their "savage" past as caravan raiders to their current status as enlightened and modern people. This adoption of a negative
image of the past makes sense in the current economic, political, and
"Cultural"[1] context of Pakistan in which Hunza is situated.

I argue that this narrative of past caravan raiding engages two powerful
discourses that have been worked and reworked in the construction of
Hunza as a remote area in the domain of Pakistani geopolitical nationalism:
the discourse of primordialist nationalism (Anderson 1991, 11) and the discourse of development. In both, Hunza emerges as a remote area in new
ways. Hunza's marginality from mainstream Pakistani culture, ironically,
makes it attractive for an official nationalist discourse in which the region's
primordial ties to the ancient Indus Valley Civilization are invoked, giving
Pakistan historical depth. Hunza is appropriated in the official Pakistani
nationalist discourse as part of the ancient Pakistani heritage linked to its
pre-Islamic past. The nationalist discourse celebrates Hunza's remoteness
as part of Pakistan's commitment to a more tolerant and pluralistic yet
Islamic national identity in which ethnic and religious minorities have a
comfortable place. In this discourse, Hunza's caravan-raiding past becomes
an important marker of antiquity and also a cultural resource, a touristic
attraction: through Hunza's ancient past Pakistan projects its own venerable
past, and presents the people of Hunza now as responsible citizens of the
Pakistani state.

In the development discourse, Hunza is constructed as an economically
marginal region because of its geographical distance from centers of political and economic power, its minority ethnic status, and its harsh climate.
The economic development discourse is pushed by the global Ismaili development institutions (GIDI), a set of nongovernment organizations and
foundations working for the social, cultural, and economic uplift of the
Ismaili community scattered throughout the world. The GIDI are spear-
headed by the central authority of the Aga Khan, who heads the various
institutions of the Aga Khan Development Network. The Pakistani state
and the GIDI work together in complementary ways. The modernist and
enlightened focus of GIDI development discourse allows the Pakistan state
to claim its official nationalist discourse regarding the people of Hunza as
liberal and modernist.

The reach of both the Pakistani state and the GIDI was made possible by the construction of the Karakorum Highway (KKH), completed in 1978 and opened to foreigners in 1984, which connects Pakistan and China and passes directly through Hunza. The KKH was built in the context of Pakistan's shifting geopolitical considerations and its improving relations with China between 1960 and the 1970s. Although the opening of the area by road has reduced Hunza's physical remoteness significantly, its conceptual remoteness, ironically, has been maintained and even enhanced through its discursive integration into the Pakistani state and the global Ismaili development network.

The Karakorum Highway—Busting Remoteness

The amicable resolution in 1963 of the long-standing and seemingly intractable issue of border demarcation between China and Pakistan, a legacy from the British colonial period of the early twentieth century, ushered in a new era of good relations between the two countries that has come to form one of the most important features of the latter's foreign policy (Rizvi 1971). At the conclusion of the border treaty, the two countries agreed to construct the "Friendship Highway" (Kreutzmann 1991, 723). After the 1965 war with India, Pakistan sought, for strategic reasons, to strengthen its relationship with China and began to make this Friendship Highway a reality. The road link also provided China with the prospect of accessing the Indian Ocean for global trade through a much-shortened route via mainland Pakistan. The construction of the road linked Hunza with the rest of Pakistan, in some ways completing the unfinished colonial project of reorientation and integration of Hunza toward the political powers to the south (Haines 2004).

The Friendship Highway, or the Karakorum Highway, as it is now known, started as a joint venture between Pakistan and China in 1966 and took twelve years to finish. The two-lane asphalt road was built on preexisting jeep roads and paths that had been used by the locals for transport and travel. The road starts in the Mansehra district of the Khyber-Pakhtunkhwa (KPK) province (formerly the North West Frontier province) and runs through the heart of Gilgit-Baltistan (GB, formerly known as the Northern Areas), first in the Indus-Kohistan valley and later the Gilgit and Hunza river valleys, crossing over the Pakistan-China border at the 15,600-foot Khunjerab Pass and ending in the old caravan town of Kashghar on the legendary Silk Road in the Sinkiang region of China. The road is more than 1,150 kilometers long and cuts through spectacular mountain scenery. The total length of the KKH in Pakistan is about 800 kilometers, of which about 200 kilometers run through Hunza. Constant improvements to the road have reduced the journey time considerably over the years, although it remains long: sixteen hours. The frequent landslides that cause the road to be temporarily closed are cleared quickly and efficiently, and there are regular buses plying the route, their price depending on the comfort and spaciousness of the bus chosen.

The KKH brought rapid socioeconomic changes to the region, including a growing local economy dependent on trade with China and also in large part on the tourist trade. Since the road opened to foreigners in 1984, the region has witnessed a steadily rising flow of international tourists. By 1995, the number of hotel beds available in Hunza had increased by a factor of one hundred from 1980 (Kreutzmann 1995, 224), and by early 2000, about twenty thousand tourists had passed through. Depending on their budgets, these tourists use either public or private transport; during the summer in particular the KKH is full of foreigners experiencing "travel on the legendary Silk Road," "following in the footsteps of Marco Polo," as savvy local travel operators market it. Both national and international tourist outfits have successfully presented the KKH as a romantic space in which tourists can fulfill their desire for a sense of discovery and for historical connection to the age of exploration. Such desire transforms the local economy, creating new business opportunities and jobs: hotels, restaurants, shops, porters, tour guides, drivers, and so on.[2]

The road that brings foreign and desired goods also "empties the villages" (Tsing 2004). Out-migration has increased since completion of the KKH. Karachi, the desired destination for many people throughout GB, is now only a couple of days away, whereas less than forty years ago it used to take three days just to reach Gilgit. The penetration of the road network into Hunza has been swift and dramatic. It was only in 1957 that the first jeep was able to reach Karimabad, the capital of Hunza, but already by 1990 not only central Hunza but also most of the more isolated villages in the area could be reached by much less rugged motorized transport (Kreutzmann 1993, 36). It was the completion of the KKH that enabled the full-scale introduction into Hunza of the Ismaili religious and social development institutions and strengthened the reach of the Pakistani state.[3]

The Hunza State after 1947

Hunza's political position has changed little in its postcolonial governance by the nation-state of Pakistan. Technically and officially, Hunza's status today is tied to the princely state of Kashmir of the colonial era, whose status in turn remains unresolved between Pakistan and India since partition. Hunza today is under the administrative control of Pakistan, but its political and legal status remain in limbo. The Pakistani nation-state considers Hunza its administrative territory that one day will become part of its sovereign territory by joining the federation, but until then it is ruled directly from Islamabad without local people enjoying the formal democratic politics of self-governance held by other provinces of the state.[4]

At the time of independence, Pakistan consisted of five provinces—Punjab, Sindh, Balochistan, the North West Frontier province (NWFP),[5] which made up West Pakistan, and East Bengal to the east of India, divided from the other provinces by sixteen hundred kilometers, which made up East Pakistan. In addition, there were also various administrative appendages of the state, including princely states such as Bahawalpur and Khairpur in the south and Chitral, Dir, and Swat in the north, which were left to govern themselves with semi-autonomous status. Pakistan had also inherited from the British the Federally Administered Tribal Areas (FATA) set up in NWFP and Balochistan, areas that also had a degree of autonomy, though the state retained the status of paramount power. These areas had a legally recognized status as specially administered regions in the first constitution of Pakistan in 1956 and then in the current constitution, which was created in 1973, and they have representation in the national parliament (Kreutzmann 1995, 218).

The case of Hunza and the Gilgit Agency is different from either the princely states or FATA. As we saw in chapter 3, the political and legal relationship between the Gilgit Agency and the Kashmir state was not a straightforward one, and the British Indian government repeatedly claimed that some districts of the Gilgit Agency, notably Hunza and Nagar, were not territorial parts of the state of Kashmir. At the time of the partition of India and the creation of Pakistan in 1947, the British, who had acquired the Gilgit Agency on a sixty-year lease from the maharaja of Kashmir in 1935, left the agency under the command of a Kashmiri governor (Lamb 1968, 35).

For two months after partition, the Hindu ruler of Kashmir, Hari Singh, remained undecided about whether to join India or Pakistan; his majority

Muslim subjects wanted to join Pakistan and the minority Hindus preferred India. In November 1947, the local people of the Gilgit Agency, backed by the local army wing, the Gilgit Scouts, the Muslim Company of Kashmiri State Forces, and tribals from the FATA and NWFP, led an uprising against the Kashmir state and started a "fight of independence" (Dani 2001). By mid-November, they had expelled the Kashmiri army and civil bureaucracy from the Gilgit Agency and, on November 16, invited the Pakistani government to take over the administration of the region. Hari Singh had by this point decided to opt for India, and the state of Kashmir was divided into two parts, with the northern part coming under Pakistani administration and the Kashmir valley, Jammu, and Ladakh districts to the south coming under Indian administration. Both countries have reciprocal claims over the territory held by the other. The issue of whether Kashmir belonged to Pakistan or India was referred to the United Nations in 1948, which called for a plebiscite in the region—which was never done by either Pakistan or India. Since then Pakistan and India have fought three inconclusive wars over it.

Between 1947 and 1974, the Pakistani government administered the Gilgit Agency much as the British had done: the state of Hunza enjoyed a semi-autonomous status, though its legal position was unclear and it had no political representation in parliament. The people of the area had hoped that upon joining Pakistan in 1947, they would be able to join the federation as a province, but the issue remained unaddressed by the government and the Gilgit Agency continued to be ruled under the draconian Frontier Crimes Regulation (FCR), an administrative and legal code that had been introduced by the British to indirectly govern areas on the imperial frontiers. The FCR was exactly as its name suggests—concerned with the suppression of crime in the frontier region—implying that in colonial official and administrative discourse, crime and frontier societies were inseparable; what was needed there was not normal "civilizing" forms of governance but rather control over crimes. Unsurprisingly, the set of rules and procedures under FCR gave very little attention to the political and civil rights of the subject population because it was accepted, implicitly, that these people did not yet have what it takes to become politically conscious and self-ruling state subjects. After 1947, under the FCR, a Pakistani political resident exercised legislative, judicial, executive, and revenue powers over the GB at the discretion of the federal government (Raman 2004, 196). The people of the GB were deprived of access to the High Court or the Supreme Court; in their place a mainly ceremonial position of Chief Courts was established

(Ali and Rehman 2001, 137). During this period, Hunza's internal matters continued to be governed by its own laws and regulations, while external affairs were handled by the Pakistani political resident.

During the 1960s, under the military regime of General Ayub Khan, there was pressure from an increasingly politicized GB population to abolish the FCR, part of a wider leftist movement in the country seeking to implement socialist reforms aimed at alleviating poverty. Many young people from GB left the area during this period to study and/or work, some going to the cities of Lahore and Rawalpindi but most to Karachi, where they were exposed directly to the leftist movements gaining ground in Pakistani cities at the time. Many of them became active supporters of the new Pakistan People's Party (PPP), which had been formed in 1967 by Zulfiqar Ali Bhutto, who had resigned in 1966 from his position as foreign minister in Ayub Khan's military government.

Ghazi Mohammad, a Wakhi man in his late sixties who now runs a hotel in Passu in the upper Hunza valley popular with budget tourists, told me about his experiences when he worked as a wage laborer in Karachi during the 1960s. While working in a chemical factory, he joined the trade union and participated in rallies against the military ruler Ayub; he recalled how he, along with many other politically active men from the GB, learned the arts of politics, oratory, and mobilization during this time. Ghazi Mohammad and other trade union members from the GB presented their demands for the abolishment of Hunza state to the PPP leadership and returned to the GB in the late 1960s to organize demonstrations against the mir in Hunza. A number of them, including Ghazi Mohammad, were jailed in Hunza for some months for these activities. By early 1974, Bhutto had decided to abolish the FCR and the princely status of the Hunza state. The people of Hunza were now free citizens, but of which state?

The historical connection of the Gilgit Agency with the former state of Kashmir was, and still is, the major sticking point preventing Hunza from achieving fully fledged membership of the state. The Pakistani government's position on the GB has been that any final or unilateral settlement of the region will give a signal to India that Pakistan has abandoned its claim over the Indian-held areas of Kashmir. This policy has remained basically unchanged to this day. In recent years, the people of the GB have tried to disassociate their fate from the fate of Kashmir, arguing that they have no cultural, ethnic, or linguistic relationship with the Kashmiri people, and that therefore the final status of the GB should be resolved separately from the solution of

Kashmir (Kreutzmann 1995, 218). As Kamran Ali, a prominent Shimshali now working as assistant editor of the major national paper, *Dawn*, wrote in an editorial on October 24, 2007, "It is extremely unfortunate for the people of the Northern Areas [as GB was called back then] that their fate and freedom have remained tied to the resolution of the Kashmir dispute."

The continued failure of the Pakistani state to respond positively to the political demands of the people of the GB has spawned an atmosphere of disenchantment and disdain, with the hope of the 1960s and early 1970s largely dissipated. Indeed, many locals have aptly described their political history as "out of the frying pan into the fire," illustrating their frustration with the Pakistani state's failure to fully implement political reforms, leaving them without any constitutional status or position to defend their economic and political rights in the region. Apart from the Bhutto era, all other periods of Pakistani political control over the region are popularly believed to be illegitimate and illegal. General Zia, in particular, is blamed for stoking ethnic and sectarian tensions in the region (Sokefeld 2005, 964); people claim that he deliberately acted to keep the people of GB from uniting against the illegal domination of the Pakistani government.

Many people complain that they are unjustly considered disloyal Pakistanis by the state. They protest that their commitment to Pakistan, shown by their decision to rebel against the Hindu maharajah at the time of partition in order to join Pakistan, has been unfairly ignored, particularly given the fact that other provinces of the nation, including the KPK, Balochistan, and Sindh, had large ethnic-nationalist movements opposed to the Pakistan movement at the time of partition. The people of the GB also consciously separate themselves from people of the FATA areas who, they claim, chose to remain outside the Pakistani federation of their own accord.

Despite the lack of formal political integration of the GB and Hunza into Pakistan, there has been significant integration in other aspects, some of which have been more welcome than others. The civil and military institutions of Pakistan both maintain a presence in the GB. Although the civil bureaucracy in the GB is weak due to a lack of resources, it is structured similarly to that of other provinces. More than 95 percent of bureaucrats are local. The people of the GB are almost fully incorporated into the Pakistani civil state structure. Senior- and midlevel officers frequently make official visits, or "bureaucratic pilgrimages" (Cohn 1996), to Islamabad, where they learn the arts of governance, bureaucracy, and officialdom. Employment in the government, both within and outside Hunza, is highly coveted among the locals. Local ethnic groups often gauge each other's relative strength by identifying how many members hold high government posts; these posts are used as markers of that ethnic group's progress, development, and power. For example, Kamran Ali, the editor of *Dawn*, told me in conversation that "up until recently the Wakhis were not socially mobile, but recently they have left behind the Buroshoski Ismailis of central Hunza. Wakhis are now in senior level government positions, in the Ismaili councils, and the Buroshos feel jealous."

Some other ways in which Hunza has been integrated into the wider Pakistani nation-state have been seen as more contentious, notably the imposition of the federal government education curriculum, which not only teaches a state-sponsored version of national history but teaches it in Urdu, a language that is the mother tongue of no one in GB. From the perspective of the state, this is a perfectly legitimate step toward bringing the area into the fold of national consciousness.[6] Moreover, under Zia's military regime, in the 1980s the Pakistani government commissioned a large volume called *The History of Northern Areas*, published by the National Institute of Historical and Cultural Research, that emphasizes the historical and cultural affinity of the GB with Pakistan based on their common Muslim identity (Sokefeld 2005, 968). The Islamization of local history, with its special emphasis on the struggles between the local Muslim population and the former Hindu Dogra ruler, was part of a wider policy of Zia in the context of the Soviet-Afghan war.[7]

Today, Hunza and the GB are physically linked with Islamabad and "downcountry" by road and air. There has been an air connection since the 1960s between Gilgit and Islamabad, with three daily flights, and the road between Gilgit and Islamabad, though long, arduous, and plagued with landslides, is as well maintained as possible given the landscape. In the last two decades, communication networks have improved dramatically; the emerging market towns in Hunza have a digital telephone system, and cellular telephone service has recently been made available. Recently, the Internet connection in the GB was upgraded to high speed via fiber optic. The Pakistan radio service is the most important and popular means of receiving news of the outside world, although the proliferation of the Internet has changed the region profoundly.

The physical connection of Hunza with the rest of Pakistan has not brought concomitant familiarity of Hunza to Pakistani society at large. The average Pakistani does not know about the people of Hunza, let alone about their specific history of "caravan raiding" and their role in the Great Game. That Russia was ever a threat to the northwestern part of British India, what now constitutes most of Pakistan, is not part of any official or unofficial historical discourse. Hunza and its history are unknown to many Pakistanis, who usually lump them with the Pashtuns because they live in the mountainous north in pine forests (another misconception as Hunza is extremely arid) and have fair complexions. My many informants from Hunza find it amusing to be so viewed. They told me that in big cities they are referred to as "Khan Saheb," a common way of referring to ethnic Pashtuns in the plains. To Pakistanis, the people of Hunza are a little known and misunderstood people. This is changing to some extent, as more and more educated Pakistanis, especially the younger generation, are taking the opportunity to travel to the region, but there are still not nearly as many tourists as visit popular tourist hill/mountain destinations such as Murree, Auybia, Kaghan, and preinsurgency Swat.

Hunza in the Pakistan Nationalist Discourse

Using the Indus Valley Civilization theory, the Pakistani state at different points in its history has attempted to address the problem of national integration of its various ethnic groups. The Indus Valley Civilization theory is a response to the two-nation theory, the official nationalistic narrative that was the basis for the partition of India in 1947 (Gilmartin 1998; Talbott 2005). The latter stated that because Muslims and Hindus were two separate nations by every definition, Muslims should have an autonomous homeland in the Muslim-majority areas of British India to safeguard their political, cultural, and social rights, within or without a united India. The Indus Valley Civilization theory is based on the discovery of the ruins of an ancient civilization in the Indus Valley, primarily in Pakistan and western India extending westward into Balochistan, that flourished between circa 3300 and 1700 BC, with its peak between circa 2600 and 1900 BC. The theory based on this discovery rejects the two-nation theory about Pakistan, taking the origin of Pakistani culture back into the "remote" pre-Islamic phase, and is hence a more inclusive nationalistic theory that incorporates both the Islamic and non-Islamic past of the land that is now called Pakistan. Both of these theories have been deployed by the Pakistani state to integrate Hunza into Pakistan, and the people of Hunza have in their turn used them to construct strategic cultural identity.[8]

David Miller asserts that nationality "is an identity that embodies historical continuity" from the past in which nations see themselves as a historical community (1995, 23). But connecting a nation to its past often involves connecting it to communities of people who reside in the remotest geographical part of the nation-state. This serves two purposes. First, it replaces the temporal depth of a nation with spatial depth, a modernist idea that renders remote and faraway places in space as ancient and historical communities in time. Second, it creates an all-encompassing nationalistic narrative in which even remote communities are physically, economically, socially, and politically assimilated and integrated into the nation-state.

Such considerations of remote regions as progenitors of a nation-state are often created, as Lázló Kürti (2001) suggests, by the elites at the center and not necessarily the regions' inhabitants. In his book, *The Remote Borderland: Transylvania in the Hungarian Imagination*, Kürti argues that the region of Transylvania had been central to both Hungarian and Romanian national identities. Kürti builds on Ben Anderson's (1991) and Ernest Gellner's (1983) analyses, which argue that national elites often take rural folk societies as repositories of an authentic national culture. In the same vein, we notice that the bureaucratic and political elites of Pakistan picked Hunza as a representative past of the nation, a past that did not invoke Islam as a basis of national identity. Gellner (1983) and Anderson (1991) would, however, argue that there is no primordial nation; in fact, the very concept is modern and comes into being under certain historical and political conditions. Thus, national identities are not really about the past—rather, they are an ongoing process, subject to continual reconstruction in accordance with present and future demands, even as they remain related to the past (Hearn 2000, 7–11). This is certainly true for Hunza and Pakistani nationalism. The breakup of Pakistan in 1971, when East Pakistan became independent Bangladesh, invalidated the two-nation theory, and a search for a new national identity narrative at that crucial moment made the primordialist nationalist narrative a logical choice.

The PPP had been the winning party in West Pakistan (henceforth referred to simply as Pakistan) in the 1970 election, and in the aftermath of the breakaway of East Pakistan and the formation of Bangladesh, the PPP became the first political party to attempt to engender a pluralistic political consciousness based on the principle of unity in cultural diversity (Toor 2005, 331),[9] emphasizing, for example, the common history and cultural unity of the region in an ancient Indus civilization. Osker Verkaaik states that Aitzaz Ahsan, a prominent PPP leader, "presents a historical-territorial argument according to which territory of present-day Pakistan has been a political and cultural unit since the time of the Indus Valley Civilization" (2004, 40).[10] Shifting away from the emphasis of the two-nation theory on a shared Islamic identity, the Indus Valley Civilization theory focused on a common cultural history and geography based on the Indus River and specifically the "civilizations" associated with it.

The emphasis on Indus Valley Civilization theory did not mean that only those places associated with the Indus Valley Civilization (Harrapa and Mohenjodaro) were considered as part of the deep historical territory of the nation; rather, all people considered to be the original inhabitants of the region—and not migrants, such as the *mohajjirs*—were seen as part of a national cultural unity. Some people argue that the natural watershed of the Indus River forms the ecological basis for a separate nation and state of Pakistan.

Although, as we saw in the first and second chapters, the people of Hunza were identified as the Dards, an original Aryan race subsequent to the Indus Valley Civilization, they were "ancient" enough to qualify for inclusion in the national ethnological heritage. The PPP government used the notion of Indus Valley Civilization much more loosely and broadly to include minority societies, especially those in the "remote" mountainous north of the country. These communities became worthy of official cultural protection as symbols of the country's glorious past. Peter Parkes, for example, notes that Bhutto himself visited the Kalash, an animistic group of people in the Chitral region, during the 1970s to support minor development and cultural protection projects (2000, 258). In the 1970s, this nationalist discourse opened up a new national space for diverse folk cultures, represented in the establishment of Lok Virsa (Institute for Folk Heritage, later renamed Pakistan National Heritage Museum) in 1974 (Gilmartin and Maskiell 2003, 49). The Ministry of Sports and Cultural Affairs, which is responsible for promoting Pakistan's official national image, has used Hunza variously since the 1970s to represent Pakistani culture.[11]

Although the motives behind the new PPP government's focus on diversity and federalism have been questioned, written off by some as little more than populist tactics (Talbott 2005, 229), there were important practical implications. It was in the context of tackling rising ethno-nationalism, especially in Sindh and Balochistan, and thus valorizing regional differences that the PPP government granted more autonomy to the remaining provinces than previously (Talbot 2005, 229; Shaikh 2009).[12] For Hunza and the Gilgit Agency, this meant the abolishment in 1972 of the FCR and in 1974 the abolishment of the semiautonomous state status of Hunza, despite the objection of the mir.[13]

Hunza is an ideal type of society to represent the nation's past at the national and international levels despite being undefined, hazy, and remote in the national consciousness. The "Culture" of Hunza, with its colorful clothes and ladies' hats, dried-apricot roofs, yellow poplar trees in fall, and terraced fields nestled against the backdrop of Ultar and overlooking Rakaposhi, is in itself iconic. In the 1990s, the Pakistan International Airline (PIA) chose traditional Hunza dress as the official uniform of its stewardesses. Posters of Baltit Fort in Karimabad and Rakaposhi, the twenty-five-thousand-foot-high mountain overlooking the central Hunza valley, are commonly found in Pakistani embassies and PIA offices overseas. Dioramas of traditional Hunza life and culture are on display in the new Pakistan National Heritage Museum, Lok Virsa, opened by Musharraf in Islamabad in 2005. (Musharraf had followed a similar line to the Bhutto government of the 1970s, focusing on "unity through diversity" more than religious unity). At the official level, Hunza features prominently in the Pakistan National Heritage Museum. The main page of the museum's Web site states: "Most museums in Pakistan are archaeological, which are a throwback from colonial times. The Heritage Museum is the first state museum of ethnology in Pakistan which represents the history and living traditions of the people of Pakistan both from the mainstream and the *remotest* [emphasis added] regions of the country. The location of this landmark achievement at Islamabad enriches the federal capital and adds to its attractions."

This practice of representing Hunza as a relic of a remote past continued even under Zia. In the book *History of Northern Areas of Pakistan*, Dani presents artifacts from two thousand years ago as representative of Hunza's cultural history. He uses the same tropes of remoteness colonial officers had used one hundred years before: "Sandwiched between the high peaks of Hindukush and Karakorum on the north and those of Western Himalaya on the south is the mysterious far-off land, now called Northern Areas of Pakistan, preserving hoary human traditions in association with mountain fairies. Fairy tales, and not folk tales, linger on in human mind. . . . In the back-drop of Himalaya, Karakorum and Hindukush humanity lived on in isolated valleys, cut off from the rest of the world, seeking a livelihood out of mountain hazards and only daring an occasional breakthrough across high passes during summer interlude in the year to catch a glimpse of the world beyond" (2001, 1).

It is the trope of lifting the veil, but this time the veil is lifted from a national indigenous society that is lost to Pakistani society at large but whose presence in the country's north gives depth to Pakistani claims to be a historical nation. Dani also links Hunza to the Scythians and Parthians—referencing Greek and Persian imperial history of two thousand years ago—which allows him to forcefully reject the idea of a Dard nation of Sanskrit and Greek literature, as discussed in chapter 1. Dani presents a picture of the region that is truly qualified as the "shatter zone" (van Schendel 2002), or a zone where different people escaping persecution find refuge, as described by Biddulph a century before him. Dani speculates that the region was once part of the powerful Kushan Empire and Hunza was its regional capital (2001, 135). He attributes historical continuity to the people of the region in general, positioning them as part of the great movements of tribes in the Central Asian steppes two thousand years ago. This fuzziness about the origins of the people of Hunza is common to nationalist narratives. The people of Hunza, however, trace their origins to a specific source. In the only locally written history book—*Tarekh-e-Ahd-e-Ateeq-e-Hunza*—Beg (1935) writes that the people of Hunza are the descendants of four soldiers of Alexander the Great's army. Most scholars, including the colonial chroniclers, have rejected this claim because of lack of evidence. But the story continues to be circulated in Hunza, especially by the tourist industry and some shrewd-minded gentlemen historians of Hunza.

In 2002, the National Institute of Pakistan Studies (NIPS) at state-run Quaid-e-Azam University produced a series, Languages of Northern Pakistan. The third publication in this series was on Hunza, titled *A Look at Hunza Culture*. This book, clearly written from an official Pakistani perspective, describes the people of Hunza as still in need of introduction to Pakistani society. Writing in the foreword to the book, the then director of NIPS states that the publication will "serve to introduce them [the people of Hunza] to their fellow countrymen and interested scholars and thus contribute to the mutual understanding and sympathy within the nation of Pakistan. Finally, it will serve to orient travelers who want to know more about the area they are visiting" (Wilson 2002, xiii). According to this account, both the people of Hunza and the region's geography are still unknown. In his introductory paragraph, the author, Stephen Wilson, mentions Hunza's remoteness and isolation in the past and in the present. He states: "Modern-day Hunza may have lost some of its past mystique, but it still has a special charm and uniqueness that makes it unforgettable to anyone who visits there" (1).

Within Pakistani nationalist discourse, Hunza appears in the national celebration of folk culture and diversity. The rural folk of Hunza, who were seen by visiting American and European doctors in the 1950s and 1960s as a bulwark against global capitalism and industrial agriculture and the values associated with them, become in the Pakistani official nationalist discourse of the 1970s a repository of the country's original culture and values. Hunza is presented as a living example of the state's official nationalist narrative of a benign and tolerant Pakistan, a narrative preached from bureaucratic and civil society institutions.[14] This narrative is on a head-on collision course with the reactionary nationalist discourse prevalent today, based on the common experience of powerlessness and apathy in which the country is redefining its identity as a fundamentalist Islamic nation.

Hunza in the Pakistan Development Discourse

In official and unofficial development discourses across the world, remote areas are constructed as natural candidates for intervention. Development "refers to officially sponsored and patronized efforts to improve living conditions, especially among the poor, going beyond the charity to include overall moral and material advancement" (Ludden 2000, 251). Remoteness is often constructed in the development discourse as economic marginality. For example, McWilliams describes marginality as a "remote outpost" and "isolation from centers of economic power and processes of the global market" (2007, 1113). Anna Tsing (1993) explores how remoteness and marginality are intertwined in state development projects among the mountain community of southern Kalimantan, Indonesia. Huskey and Morehouse (1992) state that remote regions are physically, economically, and politically distant from centers of wealth and power; they are culturally or ethnically diverse and sparsely settled; and they exhibit extreme limits on their autonomy, self-sufficiency, and welfare. They define the development of these regions as the overcoming of internal and external obstacles to changes in conditions associated with their remoteness (128–29). An important point they make is that remote areas "typically have limited market economies, and they are dependent on natural resource exports, government transfers, and subsistence activities. The costs of doing public and private business are high. Important decisions affecting these areas are made in distant metropolitan centers" (129). Because there is a materialistic bias in measurement of development, we notice that remote areas also become reified categories that become amenable to administrative interventions.

Remote areas are characterized in official lists, policy briefs, and government files as areas to which it is hard to bring development. Remote areas are officially named and designated; usually a hard rule—such as distance from some place identified as the seat of economic or political power or determined by the availability of certain technologies and social services—becomes the defining indicator. Many countries, developed and developing, have adopted the phrase "remote areas" as part of the official lexicon. For example, Botswana has a Remote Area Development Program (Molebatsi 2002) and the state of Bengal in India has an official list of remote areas (*Telegraph* 2012) that are targeted for special social development treatment. Even developed countries like Australia (Argent and Rolley 2000) and Canada have used the term *remote areas* to identify their distant and less developed peripheries. The underlying assumption about such areas is that they lag behind and need to catch up with the rest of society.

Alongside a nationalistic narrative in which Hunza is constructed as belonging to the remote and deep past of the newly emerging Pakistani nation-state, we also find Hunza's remoteness constructed in a different way. In this discourse Hunza is represented as a remote and underdeveloped region. This discourse is popularized in the practices and accounts of the Aga Khan Foundation (AKF) and its development network, headed by the spiritual head of the Ismaili Muslim community, known by his title, the Aga Khan.[15] While the Pakistani state tries to integrate and assimilate Hunza into mainstream Pakistani society, the AKF and its institutions make Hunza part of a global Ismaili community. Ismailis are scattered all over the world, constituting nothing more than a small religious minority group in any state. Despite this, however, Ismaili communities are directly linked with a highly centralized and hierarchical structure of religious and nonreligious institutions, called the councils, under the direct authority of the living imam, the Aga Khan.

The council structure initiated by Aga Khan the Third built upon the existing Ismaili institution of Jamat, or local congregations, worshiping together in a single Jamat Khana (house of worship). The local councils were linked to the supreme council headed by the Aga Khan through a series of regional and national councils, each looking after the affairs of the community at a different level and hence dealing with issues of social change at these strata. Through the council structure, the Aga Khan reached directly down to the Jamat through his *farmans*, or edicts, which dealt with issues of education, social welfare, economic cooperation, and gender roles and balance.

Using Anderson's (1991) notion of an "imagined community," Jonah Steinberg, in his work on the Ismaili "trans-nation," compares the monthly farmans disseminated by the Aga Khan to his "spiritual children" all over the world through the local council structure with "print capitalism" (2006, 23).[16] These farmans are read by people in Toronto, Nairobi, Geneva, and Hunza at the same time and create a sense of belonging to a larger Ismaili nation. The Ismaili collective identity is essentially nonterritorial, but it still operates in an imagined geography in which the wider community is divided into economically prosperous and economically depressed areas.

Most of the social development work of the AKF is executed in tight coordination with the council structure, albeit informally. In this context we see how the development work of the AKF becomes entangled with the communal and religious commitment of members of the Ismaili community to each other. The foundation describes its mission as "commitment to reducing rural poverty, particularly in resource-poor, degraded or remote environments" (Aga Khan Foundation n.d.) around the world. Ismaili Muslims who are citizens of affluent Western nations such as the United Kingdom, Canada, and the United States work for the institution of their imam helping their less fortunate fellow community members living in developing countries in harsh economic and social conditions. It is in this context that Hunza is constructed a remote place, marginal to development and prosperity. The remoteness of Hunza is thus a part of the Ismaili religious development narrative, presenting an opportunity to Ismailis of affluent countries to experience hardships and face struggles while helping their fellow community members achieve a better standard of life.

One of the main institutions of the AKF working in Hunza on economic development issues is the Aga Khan Rural Support Program (AKRSP). The AKRSP's mission is to work as a "catalyst for equitable and sustainable rural development in the Northern Pakistan" (Aga Khan Rural Support Program 1993, 2). I worked for the AKRSP office in Hunza's neighboring region of Baltistan, which could be considered equally remote from the perspective of a development discourse. Most of my colleagues were local Balti Shia, but there were a handful of local and nonlocal Ismailis as well. For many outsiders like me, working in development had a romantic appeal that was enhanced by the charm of doing so in a remote area. I often thought, and other outsiders agreed with me, that there was a certain touristic quality to our development profession and experience. Most often the stay of development workers in a remote area is quite short. The data-gathering techniques used, such as "rapid rural appraisals" and "participatory rural appraisals," are testaments to this fact. Development workers themselves usually reside in large urban cities—for them, the remote area is the "field." The further the field is from the office, the more authoritative are the claims of development workers. We notice that the claim to knowledge of Hunza as a remote area was also a source of bureaucratic capital and status for the colonial officers of the Raj.

The official discourse of AKRSP divides communities within the former state of Hunza into remote and central areas. This construction has historical roots in the Hunza state's ethnic makeup and is further aided by the building of the KKH, which redefined "remoteness" in the region. The northern part of Hunza, known as Gojal, which is at a higher altitude than central Hunza and inhabited by the Wakhi ethnic group, is today considered remote compared to the southern part and, as such, worthy of special development intervention. In particular, difficult-to-reach villages like Shimshal, inhabited by the Wakhi ethnic minority, where I conducted part of my ethnographic fieldwork, are considered especially remote in official AKRSP discourse. Shimshal was not only a target of special development attention, it was also a place where the Ismaili councils experimented with an innovative governance model.

I was told by Munawar from Shimshal, a former colleague at AKRSP and now a friend and informant, that in 2000 the Aga Khan suggested that a woman should be considered as president of a local council, asking which council would be prepared to try out the idea. Shimshal, the most marginal and remote community, agreed to do so; Fatima, a university-educated young woman, was selected and remains to date president of Shimshal's local council. Fatima's older sister is a yak herder who goes up to the pastures in the summer.

Many people from all over Hunza have found work in AKRSP, but although both Buroshos and Wakhis as well as other ethnic groups from the region are employed by AKRSP, Wakhis have taken particular advantage of the opportunities available. Munawar, my Shimshali friend, now works in a senior position with AKRSP in the Gilgit office, having climbed up the ranks from the position of junior intern. He told me that for him and many other Wakhis, the AKRSP has provided the opportunity for upward mobility that was denied to them because of their marginal position in the Hunza society during the era of the Hunza state. They were able to take advantage of the educational institutions opened by the AKF and of relatively easy access to higher education in Gilgit and the larger cities of the south, given the opening of the KKH. After graduating from college, Munawar joined the AKRSP office in 1992 in Gilgit, where he was joined by a handful of Wakhis from other villages in Gojal.[17]

Farida, a university-educated Burosho woman from Karimabad, told me that the reason that Wakhis today have progressed ahead of the Buroshos is because of their belief in the work of the Aga Khan Rural Support Program as their religious duty. She argued that people in remoter areas, such as Shimshal, tend to be more religious minded, and when the AKF started working in the area they participated in its activities wholeheartedly as a matter of religious conviction. Farida's explanation had a hint of the ethnic bias the Buroshos of Hunza have traditionally harbored for the Wakhis. By claiming that they were more "religious minded," Farida meant they were somewhat simpleminded, unable to question authority or make decisions for themselves. This depiction of the ethnic Wakhi of Shimshal by the Burosho of central Hunza was heavily tainted by the development discourse that associated remoteness with marginality, simplicity, and lack of sophistication. The presence of this conceptual and subtle internal division within Hunza between remote and non-remote areas was made evident to me in a remark made by the same Burosho man who persisted in his belief that in the past the people of Hunza were indeed caravan raiders. After our discussion (with which I opened this chapter), he asked me where I was going next. I told him that I was going to Shimshal. He chuckled and said, "In the past only criminals[18] used to go to Shimshal; these days it is mainly anthropologists who go there." For my Burosho informant, the identity of Shimshal as a remote outpost of the former Hunza state had evolved from a penal colony into an anthropological attraction, just as the identity of the people of Hunza had undergone a transformation from caravan raiders to modern, rational people.

Conclusion

Almost a century and a half ago, in the 1870s, the British searched for the original Aryan people and a golden civilization in the confines of Hunza. The construction of Hunza people then as the original Aryan race was made within a unilineal evolutionary discourse in which the British saw their own past in less civilized and primitive "races." We notice similar evolutionary tendencies at play in the construction of Hunza as a relic of the past in the discourses of nationalism and development. In the former, certain ethnicities, preferably those living in remote areas and associated with remote mythologies, in this case the Indus Valley Civilization, become the progenitors of a contemporary nation-state. Their remoteness gives legitimacy and stability of time-depth and geographical continuity to the claim of nationhood as a natural and organic phenomenon. In the case of the latter (development) certain "mental" conditions and features of geographical positioning become ideal candidates for development intervention. Both of these discourses use remoteness to address issues of political modernity and economic modernization. The remoteness of Hunza, which was earlier used to talk about issues of civilization, progress, and evolution, is resignified, imbued with new meaning and put to new uses. The people of Hunza continue to engage the discourses in which their remoteness is perpetuated (nationalism and development) by strategically articulating and rearticulating elements from their historical representation (caravan raiding) and their serendipitous yet contingent geographical position (Indus Valley Civilization). But how do these people think about remoteness from their own perspective, a perspective from which *they* see other locations and people as remote, and how does this shape their social practices, rituals, and livelihoods? I will turn to this question in the next chapter, examining Shimshal, a village in northern Hunza, as a microcosm of this analytical inquiry.

CHAPTER 6

On the Edge of the World

In this chapter, I explore the indigenous notion of remoteness as it was articulated to me by Shimshalis and as I observed it during my fieldwork. I briefly look at how Shimshalis construct and experience a perspective of their own remoteness. A major part of Shimshalis' social experience with the outside world is defined by their geographical remoteness, of which they are acutely aware. I also briefly describe this remoteness from an outsider's perspective by elaborating on my own experience of being in the village. I then look in detail at the Shimshalis' perspective on internal remoteness in the context of their transhumance[1] cycle of migration and the rituals associated with it, as they roam over a vast network of high-altitude pastures around the Shimshal Pass near Pakistan's border with China.[2] Paths to the summer pastures offer considerable friction of distance due to the craggy mountain terrain. Shimshalis ease the friction of distance to the pastures by funding communal work.

To the Shimshalis, pastures are sociologically a different world. They are devoid of humans but are not empty places. Shimshalis still practice some animistic belief in which they ascribe certain humanlike qualities to the natural world, especially animals (Descola 2013, 129–30). The pastures represent a parallel world called *mergich*, which can be occupied temporarily and must be abandoned at the end of each season. Habitation in this realm is made possible by carrying out a series of rituals, called *mergichig*, which are all geared toward filling in the empty places with humans and their livestock. Shimshalis believe that the pasture world is controlled by the *mergichon*, the spirit master who also controls the wild ungulates, especially ibex and blue sheep. Shimshalis believe wild ungulates are the mergichon's livestock, just as domestic ungulates are theirs. The ritual of taking occupation of the pastures at the beginning of each season involves differentiating their livestock from those of the mergichon.

Shimshali agropastoral ceremonies are geared toward symbolic acts of claiming and de-claiming the space and territory surrounding the village. Specific ritualistic aspects show a structural element in which seasonally abandoned territories and space are claimed by announcing their arrival. This suggests that Shimshalis divide different agroecological zones into different cultural zones marked by specific materiality and morality. This awareness is reflected in the emphasis on different norms of behavior and actions in the different zones.

Distant summer pastures, although made familiar and habitable, create a temporary separation between the sexes as the women, who are responsible for the summer grazing of yaks and goats, spend the entire summer in pastures away from the village.[3] Most Shimshali men, young and old, regularly visit their female relatives and spouses during the summer months to reduce the anxiety of separation. During the winter, all smaller livestock are brought back to the village, while yaks are taken by a group of nine men, called *shpuns*, to winter pastures further away from the summer pastures. Compared to summer pastures, distant winter pastures are an alien and strange world to most Shimshalis, men and women, as these areas remain cut off from Shimshal for about six months. For about the same time in winter, Shimshal itself remains cut off from the outside world. Thus, the summer and winter seasons in Shimshal are marked by both separation within the community and isolation from the outside world. Both isolation and separation form the underlying structure of many rituals associated with the transhumance economy and collective social work in Shimshal and reflect how remoteness is constructed and negotiated culturally.

On the Road

The village of Shimshal is located in the northwestern Karakorum Mountains in the northeastern part of the former state of Hunza, or Gojal, region and is a Wakhi-speaking community. The Wakhi people in current-day Pakistan came from the Wakhan, a part of the Pamir Mountains of Afghanistan and Tajikistan, where they are believed to have been since

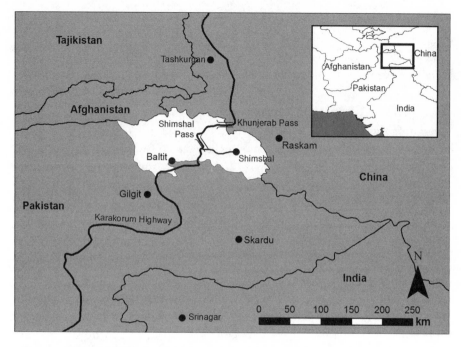

Figure 5: Map of Shimshal (map drawn by Chris Milan)

"before the time of Zoroaster" (Shahrani 2002, 45), and their language is an Indo-Iranian dialect (46). Until 1883, Wakhan was an autonomous principality on both banks of the river Oxus, ruled by a mir and populated by both the Wakhi, who were sedentary mountain farmers, and Kirghiz nomads. As the Kirghiz grew in power and the power of the Wakhis declined in the nineteenth century, a group of them fled to northern Hunza, Gojal, where they were given marginal land in the northern uplands by the rulers of Hunza, ostensibly to protect them from the depredation of the powerful Kirghiz. As we will see later in the chapter, an aspect of this history is present in the origin story/myth of Shimshal.

Until very recently, Shimshal was accessible only by a three-day trek from the Karakorum Highway. In 2003, a jeep road linking the valley to the KKH was finally completed, funded jointly by AKRSP, the government, and the Shimshalis themselves. The road, running along the Shimshal River, is one of the most dangerous roads that I have traveled on and by far offers the most "friction." The gravel road is barely wide enough for a jeep and is frequently blocked because of landslides and flash floods. I have traveled on this road in all kinds of vehicles and under all kinds of conditions. In ideal conditions an uninterrupted jeep journey from Passu on the KKH to Shimshal village takes about three hours, while riding in a tractor takes about six. My journeys have always taken much, much longer, and I have only twice completed it in a single vehicle (all other times, landslides, broken bridges, floods, and so on have forced me to change vehicles mid-journey). Despite the unreliability of the new road, it is much less taxing physically than making the journey by foot. Shimshalis often recall the extreme difficulties involved in reaching the village from the KKH in the past, an experience well known to travelers almost a century ago. G. K. Cockerill, a British surveyor, wrote this in 1922 about the path to Shimshal: "The path up the Karun Pass descends at a terribly severe gradient for 1500 feet ... here I deciphered an inscription written in Persian on the cliff, a couplet from Saadi, cursing the difficulties of the road" (1922, 103).

My first visit to Shimshal, in July 2004, was a kind of baptism by fire, befitting Shimshal's reputation as a hard-to-get-to place in the middle of nowhere. It took me two days to travel the hundred miles from Karimabad, the former seat of the rulers of Hunza, now a bustling tourist town on the main KKH. The Shimshal road had already been officially declared completed, but as with other development schemes in the region, which tend to be declared completed before they actually are, mainly to accommodate the visit of a high government official who can unveil the plaque or cut a ribbon while in the remote region, the Shimshal road was far from complete. Several of the eight bridges that span the dirt road connecting the Shimshal road with the KKH were still under construction, and many sections of the road were missing retaining walls.

I hired a jeep in Karimabad and in the afternoon the driver and I drove for an hour northward on the KKH. We turned right off the highway and crossed the bridge over the Hunza River, coming to a dirt road on the other side. This road leads to Shimshal, and Shimshal only—there it ends. Shimshal is thus a cul-de-sac. There are trekking routes one can take from Shimshal that lead to different valleys and lesser peaks, but there is only one road for motorized transport, which leads slowly into the heart of the Karakorum mountains.

After about an hour and a half further travel, during which we did not come across a single soul, we reached the river crossing of Wien-e-ban, where there is a small hut for travelers to rest in. Here, a suspension bridge was still under construction, not ready for use by motor vehicles. On the other side of the bridge, a jeep was parked, without the driver but with the key left in the ignition. With only one way to go—toward Shimshal—the jeep had been deliberately left there by a road contractor so that stranded travelers could drive it to the next stopping place at Ziarat, where they would pay for the use of the vehicle. Thus the contractor could help people traveling on the road and make extra cash. I loaded my luggage in the contractor's jeep, paid for the use of the rented jeep from Karimabad, and drove off to Ziarat.

It took me a few minutes to get used to the mechanics of my new ride, but soon I was cruising along at a breakneck speed of twenty-four kilometers per hour on the stony road. Jagged cliffs rose up to fifteen thousand feet on either side of the road and the only other living thing I encountered was an occasional Himalayan chough that glided parallel with my jeep for a while, as if giving a lonely traveler ceremonial company, and then drifted away. I arrived at Ziarat just as darkness was falling. Here was a sign of human presence: a small room, a teahouse, built of stone on the side of a huge cliff. In it, I found a group of road workers from Nagar, Hunza, and Shimshal chatting, smoking cheap tobacco and sipping salted tea. They all greeted me and I felt relaxed right away. I sat down and did not feel obliged to introduce myself; they, too, did not seem interested in the formalities of introduction. A fat man asked for the money for the jeep ride, which I paid. The proprietor was Ahmed Ullah, a man of small stature, who prepared a simple meal of *daal* and *chapatti* (lentils and unleavened bread) for me and indicated a corner of the room where I could sleep. Ahmed Ullah was a retired army *subedaar* (a noncommissioned officer historically considered the equivalent to the rank of lieutenant in the British Army) from Shimshal. He was an enterprising man of fifty-five and a world-class climber; he was the first Pakistani to successfully climb Nanga Parbat, the twenty-six-thousand-foot peak near Gilgit known as the killer mountain. In the 1980s, he had set up a small camping ground for tourists in his apricot orchards in Shimshal, where he charged a nominal rent for them to pitch their tents. He would also provide porters, guides, and food to the tourists. The Dastghil Sar Inn, as the campground was rather grandly called, also kept a guest book in which foreign tourists recorded their experiences about Shimshal. Recently, Ahmed Ullah had recognized that the Ziarat Teahouse offered a better business opportunity and he had closed down the camping ground.

The next morning, about eight young Shimshali men appeared outside the teahouse. Although my Wakhi is not fluent, I could pick up enough of

Figure 6: Summer settlements at Pamer

their conversation to realize that they were talking about the Punjabi tourist who, they had heard, had come alone to do research in Shimshal. These young men, some blond, fair skinned, and blue eyed, others with Tajik, "Aryan" features, were clad in cargo trousers and T-shirts, some wearing brand-named fleece jackets and parkas, which they probably got from secondhand clothes shops in Gilgit or were given by Western tourists as gifts. In a Pakistani city like Islamabad or Karachi, they could have easily passed for Western foreigners.[4] After a light breakfast of salted tea mixed with yak butter and dried bread, I came out of the teahouse. The young Shimshalis looked at me but did not talk to me. They did not seem shy or hostile, just slightly indifferent. As I was soaking up the warm morning sun, Ahmed Ullah came out of the teahouse after serving tea to the road workers and called to two of the young men, Karim Beg and Johur Ali, telling them to take my luggage. He told me that these two young men would look after me and take me to Pamer[5] (I had told him the evening before of my plans). Karim and Johur came forward, quietly greeted me, lifted my two rucksacks, and started walking ahead of me. The other six men grumbled and murmured a little but did not make a big fuss before quietly walking away.

As we started toward Shimshal, I asked Johur why so many of them had showed up at Ahmed Ullah's teahouse. He told me that news had reached the village (through one of the Shimshali road workers, I suppose) that a Pakistani trekker had come to Ahmed Ullah's hut at Ziarat and that he wanted to go to Pamer. These men had traveled all night to reach Ziarat in the hope of being employed in my service, such opportunities being a major source of cash income. Judging from the number of men who had arrived, I suspected there was stiff competition within the village for work with tourists as porters, guides, cooks, and so on. Karim was nineteen and Johur twenty-one; both were students in Islamabad who had come home to Shimshal for the summer to visit their families, spend time at Pamer, the summer pastures, and engage in casual work as porters with tourists. They had moved to Islamabad about eight years ago to better their education and were more attuned to popular Pakistani culture than the rest of the Wakhi.

From Ziarat to Shimshal village was about a day's trek along a relatively flat trail beside the Shimshal River, with giant rocky cliffs on either side. The dirt track was wide enough for a jeep and strewn with rocks. Hardy bushes of sagebrush or artemisia grew in clumps where they could find sandy ground amid the rocky valley floor. At this time of the year, the water of the Shimshal River was raging through the gorge and was a muddy color. We crossed a giant wall of black ice on our right side, which was the snout of the Malangutti glacier. The warm dry air was now interrupted by cold icy patches.

As we walked, Karim told me that before the completion of the road, the Shimshalis saw the traveling conditions and their isolation from the outside world not only as a great inconvenience but also as a major source of shame and anxiety. "It was embarrassing to tell our friends in Islamabad and Lahore that Shimshal was the last village in Hunza not to be connected with motorized transport," Johur added. There are some villages in Hunza and Gilgit-Baltistan that are still a considerable distance from the road and can be reached only on foot, but none are as remote as Shimshal. I later noticed that most Shimshalis were acutely aware that people from central Hunza considered them backward and uncivilized people because they hauled their possessions on their backs like pack animals. "The people in Hunza, even in Gulmit [a Wakhi village on the KKH] thought of us as not very different from yaks," Johar told me with a smile. Both men said that now the road was open, more and more people came to Shimshal. There were greater opportunities for them to work for tourists, hear stories about the West, and perhaps have a chance to cultivate relationships that might come in handy one day. While expressing extreme relief and joy at finally having a road connecting the village with the KKH, both Johur and Karim mentioned that there was, nonetheless, a sense that something valuable had been lost. They said this sense was strongest among members of the older generation, who often warned the younger villagers that the road would lead to a physical and mental weakening of the people.

Indeed, I discovered later for myself that the anxiety generated by the arrival of the road varied according to age group. The older generation of Shimshalis lamented more the loss of their geographical remoteness. I was told by an elderly person in Shimshal, "We have freed ourselves from the physical burden but at the same time we have increased our mental burden." I would later find out that with the opening of the road came an extreme sense of vulnerability to outsiders' intrusion—a sense of vulnerability quite opposite to the feeling of achievement that outsiders like me had in just reaching a "remote" place like Shimshal. David Butz (2011) captures this vulnerability among the Shimshalis in his study of road narratives in Shimshal. He identified ten narrative types in local responses to the arrival of the road, but they can be broadly be categorized into two kinds, positive and negative. For example, Shimshalis believe that with the influx of visitors to Shimshal, they will have to curtail female mobility and visibility in the village in order to reduce women's contact with the outside world. This is one of the most common effects of modernization through road construction in the villages in the region. Moreover, Shimshalis believe that increased contacts with outsiders and hence increased business opportunities will lead to internal disunity within Shimshal caused by increased competition. The Shimshalis are very clear in seeing the consequences of each choice.

Both Johur and Karim raved about the Pamer and Shimshal Pass, the main summer settlement of Shimshalis, as paradise on earth. They talked fondly about the flowers, crystal lakes, and abundance of meat, milk, and cheese there. They told me that the trek to Shuwerth, the main summer pasture settlement at Pamer, was hard and took three days. They themselves had not been to Shuwerth more than a couple of times in their lives. Later on, I would meet Johur's young cousin who resided in Karachi; he told me that prior to his visit to Shimshal he had never seen a yak. I could feel a little unseasonal chill in the air as the sun went behind the mountains and the valley bottom was painted in grayish shades. Johur and Karim quickly outpaced me on the lonely trail. I walked behind them, the distance between me and them increasing at every pace, and was soon overcome with a feeling of unusual strangeness.

We arrived in the village as the sun's last rays were dying on the eastern edge of But-but-rao, the mountaintop where the geese cross on their long winter journey to the warm waters of South Asia. But-but-rao is a major landscape feature in the Shimshali annual calendar, which is marked by the changing seasons. Johur and Karim took my luggage to an L-shaped hotel, still under construction, owned by Hamid Shah, nephew of Ahmed Ullah and himself a rising mountaineering star. Hamid Shah was a jolly-looking man of about thirty-five who ran an expedition equipment rental shop and trekking company in Rawalpindi, where he spent most of his time. Hamid Shah was the only child of relatively wealthy parents who died when he was very young. His paternal uncle, Ahmed Ullah, raised him and prevented his property from falling into the hands of certain powerful clan members. When Hamid Shah grew up, Ahmed Ullah arranged for him to marry a local girl and gave him the land on which Hamid was now building Shimshal's first hotel, the Shimshal Tourist Lodge (STL).

There were six rooms in the hotel, only two of which were completed. The hotel was located on the edge of the village at the end of the road, and thus situated the closest that motorized vehicles could come to the village. My room was sparsely furnished, with two mattresses on the floor covered with plush, gaudy Chinese blankets, probably brought by someone in the village involved in cross-border trade with China, and pillows covered with cheap embroidered covers from the south of Pakistan. The floor was covered with a yellow synthetic carpet and the window that overlooked the western end of the valley was without curtains. I would spend many days in the room, looking through this window at the road coming in and leading out of Shimshal for new arrivals in the village. In the corner of the room were a folding table and two plastic chairs and on the table was a vase of plastic roses. Everything was new and still had that "new-car" smell of petroleum and plastic. There was an erratic supply of electricity in the village, courtesy of a small micro-hydel[6] installed by AKRSP some ten years previously. Though fully fitted with bath, toilet, and sink, the bathroom had no running water; Hamid told me that a water supply scheme funded by the AKHS (Aga Khan Health Service, a sister organization of the AKRSP) was under construction and that there would soon be running water in the bathroom (and the village). He proudly told me that he wanted to give foreign tourists all the facilities that they would expect in hotels in a city. After a dinner of watery spinach, boiled potatoes, and chapatti in my room, Hamid's seven-year-old son and eight-year-old daughter brought me extra-sweet wild thyme tea and some dried apricots covered with grit and goat, or perhaps yak, hair. I then settled down for my first night in Shimshal; this room would be my home for the next few months and for every visit I made to Shimshal during the next two years.

The Village in the Middle of Nowhere

The Shimshal valley runs from east to west, and the village settlements are situated on a series of alluvial fans that descend from the mountains that bound the valley on its southern side. The total area of Shimshal valley is about 15 square kilometers. But if one includes its vast pastures, it is about 2,700 square kilometers, with a total population of about 1,100 divided among 120 households. There are now four main settlements in the village. Farmanabad is located on the edge of the main Shimshal valley away from the cluster of the three main settlements of the village, Khizerabad, Center Shimshal, and Aaeenabad, which are divided by two glacial streams. Center Shimshal was the original settlement, with the other two developing over the last fifty years as increasing population pressure led to more and more land being colonized for habitation and cultivation. All houses were flat roofed single storied, built in a traditional Hunza and Wakhi style, although more and more homes were now being built to a "modern" design.[7]

The topography of the village and its surrounds is marked by a permanent glacial zone starting at eighteen thousand feet and reaching up to twenty-five thousand in height to arid but highly sedimented valleys located at about ten thousand feet. Shimshal presents a general picture of desertlike valley areas dotted with oases of permanent village settlements. Toward the southern slopes about one thousand feet above the village is the main and oldest irrigation channel, which runs in an east-west direction on the mountain face, bounding the arable land covered with vegetation. Above the irrigation channel are areas of scree slope that rise up to fourteen thousand feet, eventually turning into broken and craggy cliff face dotted with clumps of vegetation, including wild juniper and birch. Because of the arid climate, Shimshal receives fewer than four inches of rain annually; however, the mountaintops surrounding Shimshal receive more than forty inches of snow annually.

On both the northern and southern sides of the settlement, rocky cliffs rise up to twenty thousand feet. The Shimshal River that rises in the "Pamer" at Shimshal Pass, the watershed between Central and South Asia, runs from east to west in the northern end of the valley, running along what is now a jeep road that leaves the valley from its western end. The eastern end of the valley is bounded by two large glaciers, Khurdapin and Yazghil. About five miles east of the village settlement, the Khurdapin glacier, which comes down from the east, meets the Yazghil glacier, which comes down from the south, at right angles. These two glaciers are the Shimshal River's main source of water. In the past the movement of the glaciers resulted in the periodic formation of a lake, which caused a flood each time it burst its banks. The last such lake burst occurred in 1963, during which the main settlement that used to be along the Shimshal River was washed away.[8] Today all three settlements are located about thirty feet higher than the level of the river toward the southern edge of the valley.

The original land is stony, consisting mainly of granite with very little topsoil. Over the decades, however, artful management and the diversion of glacial water melt, which brings with it eroded gravel and minerals from the mountains, has added topsoil, and with the application of night soil, the land around the village has become suitable for crop production. The cultivable land available, about 150 hectares, makes Shimshal richer in terms of cultivable land per household than any other village in the region. Most of the cultivable land is devoted to the production of wheat, barley, peas, and potatoes, and Shimshal is one of the few villages in Gilgit-Baltistan that is self-sufficient in grain production. Potatoes are becoming an important source of cash income, particularly since the road was built between Shimshal and the KKH. Unlike most villages in the region, there are very few fruit, timber, and fuel-wood trees around the village. When I asked why, I was told that in the past (until 1974, when the mirdom was abolished) people were so busy providing free labor to the mir, tending his livestock and carrying out agricultural work for him, that they did not have time to raise and care for tree plantations. A rather more complete explanation, in my view, was given to me by one of the local schoolteachers; he suggested that in circumstances where the mir was appropriating a large percentage of agricultural and natural resource produce it did not make sense to raise plantations, particularly given the hardship of hauling timber on one's back to Hunza through the treacherous mountainous terrain. Since then, however, more and more people are planting trees, particularly through the support of AKRSP, which provides saplings and advice and support to interested villagers.

Most fuel needs are still met through the traditional source, sea buckthorn, which grows abundantly, mainly directly under the irrigation canal above the village settlement and on undeveloped land outside the village. Although it is considered by most villages in the region to be an inferior form of fuel, it is the most reliable source for the Shimshalis. The extremely rough work involved in cutting, storing, and in general handling the thorny bushes of sea buckthorn is the reason, so I was told, for the Shimshalis' renowned extraordinarily large hands and thick fingers. Whether the work could have resulted in a change in hand size I am not sure, but the many gnarled, hard hands I shook certainly seems to testify to the centuries-old negotiation between those hands and the thorny bushes.

In addition to subsistence and some cash income from agriculture and livestock, Shimshalis also earn income from an increasingly diverse range of off-farm economic activities. The village is a destination for serious trekkers and mountain climbers from Western countries during the summer, and some villagers earn income working as porters for these tourists. Many more, however, spend some weeks during the summer months working as high-altitude porters in other, more popular trekking and mountaineering locations in Gilgit-Baltistan such as Skardu in Baltistan, from where the climb up K2 begins. Some also work as professional guides; Shimshalis, known for their ability to work well at high altitude and for their mountaineering skills, are sought after by international mountain expedition parties. Shimshalis are quite aware of their reputation as high-altitude porters and waste no time mentioning this to a stranger.

Shimshal's remoteness was of a classical nature. There was only one shop in the village, opened mostly on demand. The owner, Ismail Beg, lived next door and customers often had to go to his house first to ask him to open the shop. One could buy the usual items at the shop, such as matches, candles, cheap biscuits and candies, kerosene oil, rope, plastic shoes, pens, pencils, batteries, and cigarettes, which were sold only to non-Shimshalis. Then there were other items that appeared in the shop irregularly, such as cans of tuna, butane canisters, down jackets, and canned vegetables. These items were seasonal and appeared mostly with the arrival of the foreign trekkers. On summer evenings, the shop served as a gathering place for village young men visiting Shimshal on a break from college. Young men also congregated on a general playing area toward the eastern end of the village near the river where they played volleyball, cricket, or soccer.

What struck me most in Shimshal during the first few days of my stay was the stark distinction between sites of human concentration and the wilderness. Viewing the village from the top of a mountain or even from the main water channel, about five hundred feet higher than the village, one noticed the scale of uninhabited space compared to the human-altered space of Shimshal. There was no gradual tapering away of the human-modified space into wild space, no transition zone between nature and culture. Rather, the shift was sudden and abrupt. Just where the last house ended, the "wilderness" and empty zone began.[9] In such a spatial environment, every human endeavor jumped out at you and every human presence was significant. A group of five men appeared as a concentrated crowd.

My first few days were full of ambivalence. I went through a roller coaster of emotions: I desperately missed the "connectivity" of the outside world but I also enjoyed the intimate company of villagers. I also missed my home and family so I experienced remoteness as homesickness. Time seemed to stop at dull moments and I wondered why anyone would ever want to live here. I failed to connect with the mental "wavelength" of many Shimshalis at first. I struggled to write anything because nothing seemed to happen here, and this feeling was not helped by the fact that summer is an unusually empty season in the village as most women are on the pastures and men are away portering. My initial experience of remoteness changed from missing home to utter boredom. But as I settled down in Shimshal, my anxieties and restlessness went away; as people became familiar to me, their activities became predictable and the general air of randomness died away. Shimshal started to emerge as a place where things happened, people lived their lives, and thus I claimed a stake in the perpetuation of this place as a point of attachment.

Pamer, Origins, and Yaks

Although Shimshalis' agricultural activities are central to their everyday lives and subsistence requirements, it is their livestock that really form the core of their subsistence and identity. More than any other people in Gilgit-Baltistan, Shimshalis are identified by outsiders as pastoralists rather than farmers. In 2006, each household owned, on average, twelve yaks and about twenty goats and sheep, with livestock products—including meat, butter, cheese, hides, wool, and carpets—being key elements of Shimshali subsistence. Shimshalis are particularly well known for their yaks; they have perhaps the largest herd of yaks in Pakistan, ranging in number close to fifteen hundred. Shimshal's enormous pasture area enables them to keep such large numbers of livestock. Although there are about two dozen pastures where Shimshalis graze their animals, the most important for them is the Pamer at Shimshal Pass. Pamer is not only central to their subsistence practices, it is also seen as the origin of their yaks and their cultural identity.

In the first few days of my first visit to Shimshal, I made the requisite visit to the unofficial chronicler of Shimshali history, Master Daud Ali, the first educated person in Shimshal, a schoolteacher at Shimshal middle school, and the person from whom most outsiders hear the history of Shimshal. I found him sitting in the apricot orchard of his house under a tree reading a book on the history of the Fatimid Empire. A man of about sixty-five, Daud Ali had a red, lined face and piercing blue eyes. He had heard of my arrival and seemed to be expecting me. I sat down with him and we started to discuss the book he was reading, our conversation quickly taking us back into the tenth century when the Ismailis built the magnificent city of Cairo and founded the famous Al-Azhar University as the center for Islamic learning and scholarship. We discussed the arts of governance, warfare, and politics. Master Daud Ali knew the main reason I had come to see him, however, and he soon turned the conversation to Shimshali history. Since then I have heard variations of this history from different people, but the story I reproduce below is the most common and accepted one.

The founder of the village of Shimshal was a man from Hunza who, about eight hundred years ago, was sent as an emissary of the mir of Hunza to Sarikol, a neighboring state in Afghanistan (now in China) toward the north of Hunza.[10] During his diplomatic tenure, this man married a local Wakhi-speaking woman named Khadija. The couple lived happily in Sarikol, where the man became the owner of a large flock of sheep and goats and took the name Mamu Shah, or Milk King. One day, diplomatic ties between Hunza and Sarikol suddenly broke down and Mamu Shah and Khadija were forced to flee.[11] They escaped under cover of night and came back to northern Hunza. Not feeling totally safe from the Sarikolis, however, the couple decided to go east toward the hitherto uncharted valley of Avgarch (today located on the KKH about forty kilometers west of Shimshal). Here the two

Figure 7: Master Daud Ali

stayed for one year, farming and tending their livestock, which flourished. A keen hunter, Mamu Shah was chasing ibex one day when he reached the top of a mountain peak he had not climbed before, Qaroon Peak. Looking eastward, he saw a lush green valley (which would later be named Ziarat); he went home and told Khadija they were moving there. They had lived at Ziarat (about twenty-five kilometers west of present-day Shimshal) for about a year when, again chasing ibex, Mamu Shah went further east and saw the valley of Shimshal. He went back home and told Khadija that he had seen another even greener and more open valley in the east and that they were moving there. Mamu Shah claimed that he would one day build a village in this new valley. While wandering in the valley one day before they had moved there, Mamu Shah saw a hole in the ground covered by a large stone. When he removed the stone, water gushed from the hole and started flowing along the remains of an old channel.[12] The couple, isolated from the rest of the world, farmed and tended their livestock in the valley, supplementing their diet with ibex and other game.

At this time, Mamu was in his fifties; because of all his wanderings he had not spent much time with Khadija, and they did not have any children.

Moreover, by this time Khadija had become very upset with Mamu Shah because of all the trouble he had caused her by constantly wandering and moving from place to place in the wilderness, and finally settling in this god-forsaken place. So distressed was she at living in such isolation and wilderness that she stopped talking to Mamu Shah; whenever she was obliged to address him she would call him Mamu Singh—Bad Milk—instead of Mamu Shah—Milk King. One day when Mamu Shah was working in the fields he heard Khadija calling him—"Mamu Shah, come here!" Surprised at hearing his wife suddenly calling him Mamu Shah, he came into the house and saw a stone bowl filled with milk. Khadija told him what had happened.

While Mamu Shah was in the field a *pir*, or saint,[13] named Shams had come to the couple's house from over the mountains and given the gift of the stone bowl full of milk to Khadija. After years alone with Mamu Shah, Khadija was eager to tell someone about her troubles and hardships. The saint listened to Khadija's complaints and advised her to have patience with Mamu Shah, counseling that she should allow him into her bed. He announced that Khadija would bear a son, whom the couple should name Sher, who would become the founder of a great community. He told Khadija that when Sher grew up they should send him east to explore the vast pasture at the Shimshal Pass, which he should take control of from the Kirghiz nomads.

Nine months after Shams's departure, a son, Sher, was born to Khadija and Mamu Shah. When Sher became a young man, his parents obeyed Shams's instructions, sending him eastward to look for grazing areas in the region of the Shimshal Pass, or the areas referred to by the Shimshalis as Pamer. When Sher arrived at the Shimshal Pass, six Kirghiz men on horses appeared and asked him why he had come. He explained that he was following the prophecy of a saint who had foretold that he would eventually take control of the area. The Kirghiz responded that they could not just cede control of pastures to Sher; if he was really assisted by some divine power, then he must prove this by winning a polo match against them. Sher agreed but asked how was he to compete against them when he was on foot and they were on horses. The Kirghiz produced a four-year-old female yak for Sher to ride. The polo game lasted hours, but eventually Sher defeated the Kirghiz by driving the ball over the Shimshal Pass toward Central Asia. The Kirghiz, true to their word, forfeited their rights to the pasture and rode away, leaving not only the pasture but the yak in Sher's hands.

According to Master Daud Ali, who offered his own analysis of the story, Sher won the polo match for two reasons: because he was aided by the divine intervention of Shams and because his mount was a yak, an animal better suited to the high altitude of the region than his competitors' horses. For my part, I couldn't help noticing that within this single narrative was located the origin of the pastures, the yaks, and the Shimshali people. There is a structural similarity of this origin story of Shimshal with the origin stories of larger communities: nations. As we saw in the last chapter, nation-states usually look for remote societies, both in time and space, for their own origins. The Pakistani official nationalist narrative uses the Indus Valley as the origin of the Pakistani nation. Shimshalis construct their identity using the Shimshal Pass, once a remote and primordial region, as the start of their community. The colonization of these remote pastures provided the basis for the consolidation and expansion of the village but also introduced the pangs of separation and isolation.

The Travails of Travel and the Anxiety of Separation

Shimshal's remoteness from the outside world as well as from its summer pastures means that its society puts a high premium on making travel easy and destinations accessible. This point was evident to me when I made my first long, arduous trek up to the Pamer shortly after arriving in Shimshal for the first time, in July 2004. The women were well settled in by that time, having been up there with the livestock since the end of May. I was keen both to see what pasture life was like and to experience the beauty of the Pamer, which so many Shimshalis had told me about long before I ever visited Shimshal or thought about doing anthropological research there. I was accompanied on my trip by Johur and Karim. As we left the flat bed of the Shimshal River, strewn with boulders of all shape and sizes, the two quickly moved ahead of me, leaving me about half a kilometer behind.

Walking in the mountains is an art; one must set one's pace according to the terrain and distance to be traveled. I walked behind Karim and Johur at my own pace, remembering what Ahmed Ullah had told me at the teahouse in Ziarat: "When you go to Pamer, don't try to keep pace with these boys; they will tire you out the first day and then you won't be able to walk on the second day, when you have to cover very steep ground. Tire out the mountain—don't tire yourself." We had set off at six in the morning when there was still a chill in the air; by eight o'clock the first rays of sun touched the eastern face of the mountains to our left. In another hour the valley floor was bathed in sunlight and the sandy soil of the riverbed had warmed up. At eleven o'clock, after a relatively steep climb that lasted an hour, we reached Gar-e-Sar, the first "stage"[14] of our seven-stage journey. Gar-e-Sar was a steep middle phase of a mountain that overlooked the Yazghil glacier, caught in a flowing motion on the eastern side of the valley. Gar-e-Sar was the entry into Pamer-e-Tang, a narrow gorge where there is no water source and the trail runs at a stupendous height. There was a small stone hut with a hearth and a sheepskin-covered earth floor. Similar huts were built at the end of every stage, and all were equipped with basic cooking utensils. Karim and Johur, who had arrived some time earlier, had already prepared tea for me.

The next stage of the trek introduced me to how the Shimshalis have culturally marked the physical landscape around them to overcome unfamiliarity and the travails of travel. After a meal of tea, dried bread, and yak butter, we left at about twelve noon for Pust Furzin, the second stage of our journey, situated in a deep chasm amid a few clumps of wild willows. The path to Pust Furzin took us through Pamer-e-Tang, or the narrow Pamer, and was literally carved out of the cliff face. As we followed the trail, which wound snakelike along the cliff face, climbing up, descending down, and flattening at times, at many places supported by retaining walls built of large juniper logs and boulders, Karim and Johur purposefully kept pace with me. They told me that the path from Gar-e-Sar to Pust Furzin was divided into four sections—Purik ka Pareean, Zainul ka Pareean, Ata ka Pareean, and Sow ka Pareean. Pareean, as they explained, referred to those sections of the path supported by retaining walls; a pareean thus implied human agency and was differentiated from a fiduk, or a naturally occurring path. It was in this context that I began to understand the meaning of the linguistic element *ka*. It is an Urdu word that means "of," while the other words are names. So Zainul ka Pareean, for example, means "Path of Zainul." These *pareeans*[15] were found all the way up to the Pamer, each pareean named after someone who had died, generally people from wealthy or otherwise influential families in the past. Karim explained that it was a great honor to have a section of the path built or rebuilt in one's name by one's family members (they would normally pay for the work rather than carry it out themselves). Having the pareean built in one's name gives the deceased person *sawab*, or grace. In this way Shimshalis inscribe the landscape through hard work with the memory and identity of their honored deceased. Moreover, assuming responsibility for the upkeep of the pareean means that the family is still strong and has the financial resources to do so. The huts at the end of each stage are similarly constructed by different households to provide sawab to their deceased relatives; those families continue to maintain the huts and provide the cooking utensils for communal use. As some sections of the trail—and likewise the huts—are better maintained than others, Shimshalis joke that the least well maintained ones might end up ensuring that the deceased is cursed rather than given sawab!

These pareeans function as a marker of a family's wealth and status in the village and to some extent can be compared to aspects of the institutions of kula as described by Malinowski or potlatch as described by Boas. Keeping their particular sections of the track in good maintenance gives a family prestige and honor and symbolizes its wealth. These pareeans also represent the occupation and marking of the intermediate zone between the village and the pasture and a collective effort of the community to ease the travails of travel on the rather treacherous and long trek between them.

Another example of Shimshalis' self-awareness about their remoteness and eagerness to overcome it is their bestowal of respect and status on those who find new routes or tracks to old destinations. The most famous pathfinders are hunters. The path to the Shimshal Pass through Pamer-e-Tang, the one we were taking, was discovered by Mohammad Shafa, a famous hunter who in pursuit of his quarry took the risk of entering the dreaded gorge. As a result of his discovery, the travel time to Pamer has been substantially shortened: some Shimshalis, unaccompanied by their livestock, can do a return journey in one day. The alternative path is at least three days long and requires crossing two passes over fourteen thousand feet. As a result of his discovery, Mohammad Shafa was exempt from all communal work. The importance of pathfinding has increased with the arrival of mountaineering in the region. The headman of the village is an avid climber, hunter, and pathfinder. He is known to have found a daylong track to Sher-e-lik, the last stage of the winter pasture, from Shujerab, a journey that before took seven days.

Thus, making travel easier by finding new routes that shorten distances and keeping the link between the village and pasture by maintaining pareeans are important concerns in Shimshalis' lives, concerns that are the direct results of the difficulty in accessibility and the vastness of the space that Shimshalis encounter in their transhumance migration cycle. The marking of these pareeans by dedicating them to ancestors addresses the isolation that Shimshalis feel in this world of rock and snow.

Another, perhaps less conscious, way in which the Shimshalis have made adjustment to norms of travels in the region is by strict adherence to timetables. For two days in our travel we did not come across a single soul. Johur and Karim did not seem to find anything unusual about this, never even mentioning it. Later on, I realized that in Shimshal adherence to time-tables for routine activities, travels, and festivals is extremely important. Each festival is set on a particular date well in advance. The same kind of exactness related to time is applied to the phenomena of the physical world; for example, I was told by everyone in the village—without a single exception and with absolute certainty—that the ducks from Central Asia arrive in Shimshal every year on October 15! This is when a flock of ducks come over the But-but-rao, the peak dominating the northern side of the valley. Sure enough, on October 15, I was taking pictures of Hamid Shah's son and his brother-in-law who had shot sitting ducks in a field about a hundred meters from my hotel room.

This strict adherence to timetables is most pronounced in the case of travel between the village and Pamer. Each person leaving Shimshal to visit Pamer leaves behind an exact itinerary, including which path he or she will take.[16] That is why Johur and Karim knew when they left the village that there was no one due on this road while we were on it. Likewise, they knew exactly how many people were up at Pamer, how long they have been there, and how long they intended to stay.

The Shimshalis explain that their exact observance of timing is for two reasons. First, the vastness of the territory over which the Shimshalis roam makes keeping track of people a necessity. Indeed, during my stay I observed that when a person failed to show up in the village on the due date, people immediately went in search of that person. This adherence to a timetable, then, addresses the real problem of dangers faced by Shimshalis on treacherous and isolated tracks. Second, the seasonal windows in Shimshal are very short, meaning that ideal times for the movement of animals and agricultural workers are limited.

Occupation of Remote Space

The journey to Pamer tested both my physical and mental strength to their limits. At times, the path was barely wide enough to fit the sole of one shoe as it passed three thousand feet above the Shimshal River winding below.[17] The days were hot and the nights cold; many times I hinted about going back to the village, but Johur and Karim's encouragement kept me going. They constantly told me, "Pamer pay puhunch kay dill khul jata hey"—Once you arrive at Pamer, your heart opens up. I held to that thought like a mantra.

On the third day of the trek, when we negotiated the gentle slopes of Gulcheen-e-washq to arrive at Abdullah Khan Maidan, I began finally to understand what Johur and Karim meant by "at Pamer your heart opens up." The opening of my heart was matched by the opening of the landscape: from the narrow, rocky gorge to the gently rolling green hills of the pasture. When we arrived at Shuwerth, the settlement just below the Shimshal Pass, we were greeted by Qurban, a retired army man of about forty-five who had come to spend some time at Shuwerth. As he greeted me he said, "You must be a good person, with a clean, clear heart, because the day you arrived the weather cleared up; it has been raining and snowing here for the last three days." I was flattered by his compliment. I pitched my tent by one of the two lakes in the pasture and fell into grateful sleep.

The next day I was woken by the earth shaking and a sound I had never heard before. I looked out of my tent in alarm to see yaks passing on either side, heading away from the settlement toward the higher pastures. There were at least four hundred in the herd that passed by, mostly females and young males interspersed with the occasional large male. They all followed the trail of the leaders and entirely ignored me, despite passing within inches of me and my tent. The yak procession lasted about twenty-five minutes, followed by four young women whistling and shouting, "Oui, oui, oui!"

After breakfast I visited Karim's family hut, as he had invited me in for tea. The space was small, dark, and smoke filled, divided between sleeping area and cooking and butter-making place. The floor of the sleeping area was covered with a yak-hair carpet, the bedding rolled against the wall, on which we leaned while seated. Much of the small space was taken up by wooden drums, some as high as four feet, used to store butter and make cheese. In one corner, Karim's sister was busy making *qurut*, the local cheese, beating down the whey with a six-foot-long, three-inch-thick wooden pole. Karim's mother, Bilqees, prepared yak butter tea for us, warming it on the fire of yak dung, and offered it along with *moleeda*, a local dish of boiled barley and fresh yak butter, and some qurut. This was the usual food offered as a sign of hospitality. Bilqees was, Karim told me, forty-seven years old, but to me she looked closer to sixty-seven. Her hands were bigger than mine, her hair dirty with soot, her eyes red, and she had a chesty, wet cough. After serving us food, she sat down on the floor in front of a large milk-filled sheepskin and started shaking it back and forth in a rhythmic motion. This was her main and most important task: to make butter. As the senior woman of the household, she did not have to go out with the yaks as the younger women did. Fresh milk from the morning was being boiled over the fire in a large pot. Bilqees saw me watching her and said to me, "It's true, life is hard up here, but it's also hard in Shimshal, and at least up here you get fresh cheese, milk, and butter!"

That afternoon at about four o'clock, the younger women and girls who had taken the yaks out for grazing in the morning began to return. There were female yaks followed by their young ones, some young males, and a few fully grown males (though most of the latter are left to wander night and day, as it is uncomfortably hot for them at the settlement—even at fifteen thousand feet—so they stay under the glacier zone at eighteen thousand feet). The milk-giving yaks were corralled in one pen, with young suckling males and females, and the fully grown males kept together in a different pen. Now began milking time, the job of the older women. As the female yaks were milked, the young male yaks in the other pen began to play, head butting each other. The place was echoing with women shouting, yaks grunting, and young boys and yaks frolicking. As the milking and feeding were completed, night began to fall and activity within the pens and inside the huts began to slow down. As the sun went behind the mountain, the temperature suddenly dropped, and fires of yak dung were lighted inside every stone hut. I lay down in my tent. Only the occasional grunt of a male yak broke the deep blanket of silence.

During the next few days, I spent my time walking in the pastures and visiting some of the side valleys. Master Daud Ali had offered to show me some of the important historical sites that had been referred to in the story he told me. He took me to Qoeegar, where Sher was supposed to have met the Kirghiz and played the game of polo, about five kilometers down the Central Asian side of the Shimshal Pass. A little way beyond this was a grassy flat piece of land, Touchighail, where there were still remains of an old Kirghiz camping place. Just below the main settlement at Shuwerth was Valio, where anyone who dies while on the pasture is buried. Many of the graves here are adorned with blue sheep horns, testifying to the prowess of the Shimshali hunters and the availability of game. Facing toward the Central Asian side of the Shimshal Pass, a rocky cliff rising up some seventeen thousand feet is Sher Rock, named after the founder of Pamer. From the narrow valley trails of the Pamer-e-Tang to the grassy meadow of Qoeegar to the high point of Sher Rock there were signs of Shimshali physical and mental occupation.

Shimshalis' encounter with remoteness is facilitated by their yaks. It is the yaks that make the Shimshali herders traverse the two worlds of pastures and village. When on the pastures, Shimshalis follow the paths of their yaks and in the process go to places where no other human would go. Shimshali herders told me that they sometimes spend days out in the area looking for lost yaks; such excursions usually become the setting within which mythical and folk stories are placed. Shimshalis often ascribe an almost human status to their yaks, representing a kind of cattle complex phenomenon observed in pastoralist communities (Herskovitz 1926). Talking about their yaks to me, Shimshalis often described them as *meherban*, that is, compassionate or generous. Shimshalis believe that some yaks are too strong headed and want to live without any human supervision and thus constantly try to run away; Johur had a white yak that had a reputation for running away repeatedly. Indeed, making comparisons between animal behavior and human behavior is perhaps one of the most long-standing human cultural activities. Johur, describing his previously tense relationship with his brother-in-law, told me, "He was bit like a yak; perhaps he had been a shpun too many times."

Shimshalis see themselves and yaks as inseparable. Indeed, Shimshali herders, both men and women, are aware of the fact that when they are on high pastures, moving with their yaks, they are in a structurally opposite position to the "normal" situation of sedentary farmers. This realization comes from a unique awareness on the part of the Shimshalis that it is the yaks that have domesticated them, rather than the other way around. "We are their servants and they are our masters," announced Qurban. Bilqees nodded her head in affirmation. This argument of who has domesticated whom has been advanced by scholars in other pastoral contexts, such as reindeers domesticating Eskimos or cattle domesticating the Nuers, and it is equally applicable in the context of the transhumance cycle of migration in Shimshal.[18] Scholars, however, have scarcely tested whether this insight is generally recognized by the pastoral communities themselves.

Integrating People into Pastures

In May 2007, I journeyed with the summer *kooch*[19] up to the Pamer to witness the rituals of temporary occupation of the mergich realm. Unlike my last journey to Pamer, which was through the narrow gorge of Pamer-e-Tang, the kooch takes an alternative route through Zat-Ghurban, a track suitable for yaks and other animals. When we arrived at Shujerab, the last stage before reaching Shuwerth, the shpuns had already arrived from the winter pastures and were waiting with the yaks tethered in the vundun, an open corral of stone walls about five feet high. As the members of the kooch arrived, they were offered yogurt and milk in large containers by the shpun, welcoming them gently to pasture life. It was almost dark as we were arriving, and very cold, with ice from the previous winter covering half of the settlement ground. Located at about 12,500 feet, Shujerab is a settlement of about fifty small huts, surrounded by low and rounded peaks, and its grasses become available by the end of May. It serves as the first stage of the summer migration; the women and the livestock stay there for two to three weeks until mid-June, when the animals are taken up to Pamer at the Shimshal Pass.

The next morning, the official ceremony of handing over the yaks was performed. Three elderly women representing the three main clans went toward the vundon, at the door of which stood the shpuns. The women were carrying flour and the branches of spunder, an indigenous shrub, which they offered to the shpuns as they formally asked permission to enter the vundon. The other members of the kooch sat on the side of the hill watching the proceedings. Once the three women had entered the vundun, all the other women joined them and started milking their yaks.

After the milking was finished, the door of the vundon opened. The three elderly clan women stood at the door, and as each yak rushed out, the women smeared yogurt, which was given to them by the shpuns, on its back. After all the yaks had been released, it was the turn of the sheep and goats, except for one large ram that was kept inside the corral. A small tin can filled with stones was tied around its neck, and when the other animals had gone a distance of about five hundred feet this ram was released. The ram ran toward the herd with a jingling sound that caught the attention of the other animals, which all ran toward it. The jingly ram soon disappeared within the herd, though the jingling could still be heard. Johur told me, "That means this year will be good, we will not have too much loss from predation." He explained that if the jingly ram and herd had gone in different directions, it would have indicated that the coming year would be particularly bad for predation and people would need to take extra care. For one week, Shimshali women milk the yaks at Shujerab, but they don't consume any of it. They make cheese and butter from it and start consuming fresh milk only after the animals have been on the pasture for one week.

Figure 8: Trek through Pamer-e-Tang

Shimshalis believe that once they vacate the mergich, the realm of high pasture summer settlement, in the fall, the settlements are taken over by the mergichon,[20] who live there for the winter. When the kooch returns the next year to take over their huts, members must ask the mergichon to vacate their huts and in general ask permission to use the high-altitude zone for this year's grazing. In addition to asking for permission to use the pastures during the summer months, the rituals also serve to ask the mountain spirits for their help during those months. While staying on the pastures, the Shimshalis believe that, although they have permission to stay, they are still subject to the regime of the mergichon and need their help for successful grazing and, in particular, hunting (this is particularly important at the winter pastures of Raskam, where hunting is an important source of both sustenance and pleasure for the shpun). As I have mentioned, the Shimshalis believe that the wild ungulates are the livestock of the mergichon, which they herd just as the Shimshalis herd yaks, sheep, and goats; thus, permission must be sought to hunt these animals. While I was at Shujerab, Jaffar, one of the shpuns who had spent the winter in Sherilik (the name of the settlement at the pasture of Raskam) with the yaks told me: "Just as we ride our yaks, these mergichon ride the ibex and blue sheep. They have saddle marks on them; I have killed several blue sheep while being a shpun which had these marks. We can't see the mergichon, but they are there. As a hunter you must seek their permission to hunt a blue sheep. If your heart is pure, the mergichon come to you in dream and tell you where you can find blue sheep. It has happened to me several times. It also happens to Lele Shah. Likewise, they come in our dreams and tell us which valleys to take our yaks to because there the grasses might be better."[21]

The structure of this ritual, again, creates divisions between Shimshalis and the mergichon, symbolizing their presence in an otherwise alien, not empty, land. The parallel universe of the mergichon is created by the Shimshalis to overcome the isolation of the land in which they roam. By making the pastures the realm of other beings, Shimshalis ensure that they are not alone in this vast landscape. All the rituals associated with the occupation of and departure from the Pamer discursively construct this otherness in the landscape. The jingly ram is about occupying an alien land. Likewise, the marking of yaks with yogurt is done to identify them as belonging to the humans and being different from the wild animals, such as ibex and blue sheep, that belong to the mergichon.[22] The significance of using yogurt versus, say, milk, emphasizes human agency and claims to ownership.

In this ritual, it is not the accuracy of the prediction that matters, it is the mental satisfaction of announcing their presence in the realm of the mergich, the parallel universe. This ritual also has the utilitarian role of reemphasizing the importance of vigilance against the risks of snow leopards and wolves. But its most important function is to symbolically fill the space and territory of Pamer with human signs and reclaim the land that had been left vacant at the end of the last grazing season.

Figure 9: Johur and Karim on the way to Pamer

Figure 10: Johur and Karim looking for yaks on the way to Pamer

Integrating People into Village

There are two rituals that mark the reintegration of people back into village life. The structure of the ritual suggests that Shimshalis see their pastures and village as different realms, marked by two different kinds of conditions. The physical point at which the ritual of reintegration of people into village occurs is, appropriately, a bridge located just outside the village en route to Pamer. The first ritual is the return kooch, or the winter kooch, when at the end of the summer women return to the village and are greeted at the bridge. The second ritual is the festival of the *cheneer*, an occasion celebrating the productivity of the pasture through the ritual presentation of gifts of dairy and meat products from those on the pasture to the village at the height of the summer season, and in turn gifts of agricultural produce from the village to those on the pasture. The cheneer is a caravan of produce brought down to the village and greeted at the bridge. This highly ritualized greeting marks the integration of people from the pasture back into the village.

I attended the cheneer festival on August 1, 2005. On the morning of the arrival of the cheneer, I saw groups of men and some women, all dressed in their best clothes, walking toward the bridge to welcome the caravan. I joined a group of elderly men. Earlier that morning, Master Daud Ali had harvested half of his peas and barley and sent some to his relatives up on the pasture with a group of older men whose task it was to take up such produce. I asked Master Daud Ali about the significance of the bridge, to which he replied there wasn't any: "We like to go to the bridge because it is an entrance to the entire village from Pamer, so all can gather here to receive their loved ones." Therefore, the nature-culture division I had theorized was signified by the crossing of the bridge was not imagined by the Shimshalis as such, but their actions indicated otherwise.

There was much fanfare at the bridge. About fifty households were represented, mostly by older men, who were usually responsible for looking after the crops in the village during the summer months, and a few younger men who had come from downcountry. We eagerly waited for the cheneer to arrive. At about eleven o'clock, five yaks were spotted in the distance at Band-e-Sar, the last settled area en route to Pamer, followed by three men. Here was the cheneer, led this time by Mohabat Shah, an elderly member of Ghazi clan, who had brought the produce of the pasture people to the village people. As the yak caravan came closer, excitement grew. People started guessing whose yaks were being used this year for the cheneer and who else was with Mohabat Shah. When the yaks arrived on the opposite side of the bridge, they came to a halt. From the village side of the bridge, the *numberdaar*, the traditional village head under mirdom (the position is only ceremonial these days) crossed to the other side and welcomed Mohabat Shah and his companions, pouring them a cup of tea from a thermos he carried and offering them bread. After they drank the tea and ate their bread, they began to send the cheneer yaks one by one across the bridge. Iqbal Shah, the *kamaria*, the ritual priest, of Shimshal, offered prayers and sprinkled wheat flour in the air. At this point, the men and women gathered by the bridge formed into two rows, squatting in the dry riverbed. Mohabat Shah and his two companions unloaded the containers (large fiberglass drums that foreign expeditions use to carry equipment) of qurut, butter, and meat from the yaks. They then served the hardened qurut and clarified butter to the people of the village, who had brought plates with them to the bridge. After this offering had been eaten, Mohabat Shah took meat out from one of the containers and laid it on a plastic sheet. One representative of each household walked up to Mohabat Shah, took from him a portion of meat, thanked him, and moved on. This distribution of meat went on for about an hour, after which the people slowly dispersed and went back to their houses.

The winter kooch ritual, which I witnessed in 2005, takes place as the summer season draws to a close. A new group of shpuns leaves the village to gather the yaks from the women at Pamer and take them to the winter pasture. After handing over their yaks to the shpuns, the women pack their belongings and supplies and return to the village with the sheep and goats plus a number of yaks to carry the large quantities of butter and cheese prepared during the summer. The return kooch—the procession of women and animals—is reintegrated into the village, marking the completion of the annual cycle. The kooch is received at the same bridge as the cheneer. The return kooch date was set at October 22. The evening before, news came to the village that the kooch had arrived at Band-e-Sar and would cross the bridge and officially enter the village the following day. That night, the kooch, comprising some three thousand goats and sheep, forty yaks loaded with the season's livestock products, about fifty women, and some men who had gone to the Pamer to help bring back the kooch, would camp on the other side of the bridge. All those in the village who had family members in the kooch would join them for the night, enjoying an evening of music, singing, and dancing. I sat on the porch of my hotel and waved at groups of villagers as they walked past on their way to the bridge, laden with sleeping bags and pots of food, including a dish of pancakes layered with cheese and yak butter called *chilpindoq*, which is particularly associated with this occasion. The mood was festive; the anticipation of seeing loved ones was evident from the jubilant faces of many, especially youngsters who had been separated from their mothers for five months. That night there was dancing, singing, and a general atmosphere of merriment on the other side of the bridge.

The next morning, I joined those remaining in the village at the bridge to welcome the kooch and its reintegration into the village. On the other side men and women were busy packing up their camps, loading their yaks, and gathering together all the small animals. The numberdaar, Money Beg, offered prayers and gave the signal for the crossing of the bridge to begin. The crossing lasted all day. People looked for their own animals in the herd, which they had given to their relatives or *skuewene* (subclan) members to take to the Pamer for them. Children chased their family's livestock, finally herding them back to their homes, where they would remain until the kooch left for the Pamir the following spring.

The bridge, thus, can be seen as the bridge of alterity; here the two sides of Shimshali people converge: those who lived on the pastures and became temporary strangers and those who stayed in the village and remained familiar. Finally, when the first group crosses over the bridge, the Shimshalis become one, village people again. For the Shimshalis, then, the pasture represents a remote world connected symbolically through the bridge to the familiar confines of the village.

The seasonal separation between Shimshali men and women creates temporary strangeness and remoteness in relationships, manifested in the extreme politeness and formality that one observes between couples in Shimshal in their everyday interactions. This formality is almost of a ritualistic nature. We also see formality and separation between the sexes during festivals when, despite the generally open and "liberal" gender relations in Shimshal, men and women are spatially separated.

Shimshalis' sense of remoteness is expressed through various rituals geared toward decreasing their isolation. The experience of remoteness among the Shimshalis is tied to their subsistence and changes with the seasons. While summer is marked by mobility and freedom, it is also the season of separation between the sexes and between family members. Winter is the season of integration but also of isolation in two senses: the

Figure 11: Kooch returning from Pamer

shpuns are isolated from the village and the village from the outside world. The sense of mobility and freedom or lack thereof associated with each season was described to me by my friend Munawar: "In our village we say, 'Khosh Tabistan-e-Munzilam, Sakht Zamistan-e-Durgilam'" (During summer I am flying high up, but during winter I am imprisoned inside a locked door). He stated that in winter one felt imprisoned in a rock castle from which there is absolutely no escape, whereas summer is a season of adventure and travel.[23]

Conclusion

The mutual dependence of animals and humans creates two realms for the Shimshalis, or two kinds of space and territories: one belonging to the animal world and the other to the human world. Structurally, Shimshalis acknowledge that the permanent position of humans is in the village, with their permanent fields and houses with hearths. Their presence on the pastures is temporary and in order to facilitate this transitory stay, they must propitiate the permanent forces of this realm, the invisible mergichon who own the wilderness zone and its animals. Until 2003, most of the Shimshalis' rituals were intended to address these issues of separation, isolation, and temporary occupation.

Since the arrival of the road in Shimshal in 2003, however, these aspects of Shimshali remoteness are changing. Previously, rich Shimshalis used to give their surplus animals to poorer members of the community: a redistribution of wealth. Today they prefer to sell them.[24] Ease of accessibility to the outside world is engendering a new sense of remoteness among the Shimshalis—the pastures are becoming truly alien and distant, culturally and emotionally. There are changing conceptions of space and territoriality among the Shimshali population as it becomes more sedentary and fewer and fewer people are engaged in the transhumance economy. People of the younger generation are no longer part of the separation and isolation cycle associated with the transhumance economy because they are now attracted to other livelihood opportunities outside the village made possible by the opening of the road.

The sense of remoteness emerging among the younger generation of Shimshalis is of a "modern" kind, with a touch of touristic sensibilities. As this trend is still unfolding, I will visit this point in the next chapter. With the arrival of the road, for many Shimshalis, the direction of their travels has changed; it is now in the opposite direction, away from their pastures. What was remote to those of the older generation, the outside world, has become familiar to those of the younger generation, who are becoming increasingly integrated into that world. Conversely, what was familiar, albeit separated, to the older generation, the pasture, is now being abandoned, becoming truly remote for all Shimshalis.

CHAPTER 7

Strange Strangers in the Land of Paradise

After the Karakorum Highway was opened to foreign travelers in 1984, Hunza became a popular destination for Western adventure tourists who were drawn by its magnificent mountain scenery, culture, and role in the past history of the Great Game. Tourists to Hunza include serious mountaineers and trekkers who are least interested in the people, backpackers traveling on a shoestring budget who stay in cheap hotels and regularly interact with local people, and cultural tourists who are interested in local history, religion, song and dance, cuisine, architecture, rituals, and festivals. Most tourists come from Western European countries or Japan, although recently tourists from China have also started visiting in substantial numbers.

Hunza has long been the main stopping point for visitors to the region, especially those traveling to Sinkiang in China on the western Himalayan trail starting in Indian Himalaya, then traveling to Pakistan up the Indus Valley on the KKH to reach Hunza before finally going to Kashghar in China. During the 1980s and 1990s, the number of tourists visiting Hunza increased dramatically, but as a result, the "character" of the main town in Hunza, Karimabad, took a correspondingly dramatic turn for the worse, according to both Western tourists and local people. Over the years, central Hunza, especially Karimabad, became more and more crowded, including busloads of the type of tourists not usually associated with trekking and adventure. This commercialization of tourism in Hunza fueled a desire among the more serious Western culture and nature tourists to find hitherto unvisited "authentic" villages and unexplored passes. During the mid-1980s and 1990s, Shimshal became popular with tourists looking for an "off-the-beaten-track" experience. Between 1984 and 2004 (after the opening of KKH but before the completion of the road connecting it with Shimshal), each year only a handful of Western tourists made the strenuous journey, three days' trek over rugged mountain terrain, to reach Shimshal. Shimshal became particularly attractive not only because of its remoteness but also because of the legendary hospitality of its people.

In the nineteenth century, tourism was the domain of only the upper classes of the Western industrial world, but during this time a whole set of supporting services emerged, such as vacation companies, photography equipment, and published guidebooks standardizing the sites and sights to be visited and viewed (MacCannell 1976; Urry 1990). Structural changes in Western societies also led to the creation of mass public transport systems, mainly to support economic expansion, which were then also available to transport travelers to and around European tourist destinations (Pagenstecher 2001, 2). In her seminal work, Valene Smith (1977) argues that changes in Western economies in the second half of the twentieth century, particularly the expansion of capitalism, resulted in a dramatic increase in tourism because of increased disposable income and leisure time and decreased work hours. Smith argues that although the motivations for contemporary tourism are varied and multiple, generally speaking they are related to a desire to get away from the routines and structures of modern and mundane life. Referring to the context of mid-twentieth-century Europe, Pagenstecher (2001) describes tourism as a romantic escape from the industrial world sought by those seeking images of untouched nature and untouched history—echoing the motivations of the hunters and medical travelers discussed in earlier chapters. Similarly, and referring specifically to the Northern Areas, Allan (1986) discusses how mountains have taken on particular significance to European tourists who feel that European mountains have been overcommoditized.

As tourism was democratized in the last quarter of the twentieth century, it also began to become more specialized. No longer were people interested only in traditional sightseeing activities; rather, tourism focusing on culture, environment, ethnicity, and adventure also became popular, resulting in specialization within the industry targeting specific clients. Tourism focusing on little-known places became popular in the 1980s, resulting in the further expansion of tourism and a permanent crisis in the industry as ever-new "unknown" places had to be found to feed the growing demand.

Anthropologists and geographers have noted cases similar to Hunza's throughout the developing world in the 1980s: once remote and isolated places, attractive destinations for a certain kind of tourist for their authenticity, lost their appeal because of modernization and commercialization (Dearden 1996; Zurick 1992; Adams 2005). This desire to travel to faraway places is not new, as Jas Elsner and Jean-Paul Rubies (1999) discuss in their historical view of tourism. Between 1500 and 1900, within the context of European imperial expansion, knowledge of distant lands acquired a new dimension with the emergence of a scientific and empiricist paradigm, resulting in an increase in curiosity on the one hand and a demand for ever more detailed information on the other (4).

Many studies show the heavy influence of the tourism industry in constructing and marketing some destinations as pristine Arcadias, inhabited by the "exotic," the "other": "primitive," "innocent," "simple folk."[1] Writing about the inexorable expansion of tourism into new and unvisited places, Zurick states, "The point is that adventure travelers are tourists and if, in their quest for the 'authentic experience,' they are propelled into ever more remote locations, then quite clearly the continued geographic expansion of adventure travel is inevitable. This suggests that once a place becomes too successful in the adventure circuit, it is no longer a desired destination" (1992, 614–15).

Philip Dearden (1996), writing about trekking destinations in northern Thailand, states that popular trekking routes lost their appeal to adventure tourists once they became modernized. Most social science studies of tourists' motivation to visit little-known and undiscovered sites build on the influential work of sociologist and anthropologist Dean MacCannell. He argues that international and Western tourists imagine that somewhere "in another country, in another lifestyle, in another social class, perhaps there is a genuine society" (1976, 155). MacCannell states that tourists often view the tourist industry's marking of cultural sites in remote locations as staged authenticity and they therefore become convinced that there is some "real" authenticity to be discovered behind the staged one, so they embark on a trajectory of discovery (see also Pearce and Moccardo 1986).

In this chapter, I explore tourists' representations of Shimshal, drawing mainly on their accounts written in a guest book kept at a local hotel. I show that Shimshal's remoteness is constructed in the socio-spatial domain of global tourism, which both constructs and consumes remoteness as a condition for its own perpetuation. I particularly look at how Western

tourists represented Shimshal as a backstage authentic society, isolated
and remote from the perils and corrupting influences of modernization.
In this way, we notice that Shimshal's remoteness, constructed within
the recent discourse of authenticity, resonates both with the discourse of
the nineteenth-century British explorers and Great Gamers who came to
the region looking for the original Aryan race and a mythical golden civilization, and with the discourse of the mid-twentieth-century health tourists to
Hunza who came in search of an ideal rural society, a Shangri-la.

In this guest book, tourists have not only registered their traveling
experiences and views of the local culture but also debated how the completion of the road, bringing increased contact between Shimshalis and tourists, might change Shimshali society. A recurring feature mentioned by the
tourists about their experience in Shimshal is the local tradition of hospitality. An overwhelming majority of the entries described being invited by
random Shimshalis into their homes. What attracted and fascinated these
Westerners about Shimshal was not only its geographical remoteness from
modernity but also the hospitality of its people to strangers—and they increasingly associated the latter as a result of the former, expressing their
fears that increased disturbance from the outside world would put an end
to this tradition of hospitality. The entries show also that Shimshalis
participate in their own construction as a remote society.

After analyzing the entries made in the guest book and using oral history
to examine what produced that culture of hospitality, I look at the ways in
which Shimshalis today have responded to tourists' desire to experience an
authentic hospitable culture. I use the specific example of a hotelier in
Shimshal who is building a "traditional house" to cater to his guests' desire for
an authentically "hospitable" Shimshali experience. I examine how Shimshal
is being described as remote in a new way, in the discourse of the tourism industry, and how this is changing the definition of hospitality. The changes in
domestic space and functions as a result of this modification show that the
Shimshali house, as an ideal structure and design, is at least symbolically under stress from the tourists' expectations and demands for authenticity.

Karimabad: A Spoiled Paradise

Before delving into the representation of Shimshal made by tourists between 1984 and 2004, it is worth first describing the scene in
Karimabad in central Hunza, which has increasingly become the focus of a

certain class of tourists' anxiety about what might happen in Shimshal. After the opening of the KKH in 1984, the numbers of tourists visiting Hunza increased dramatically. In the 1970s, fewer than one hundred tourists went to Hunza per year, but by 1996, twenty thousand tourists were passing through Hunza annually. Karimabad, the main town of central Hunza, presents a picturesque maze of streets winding through terraced fields clinging to a steep mountainside, with the dramatic centerpiece of the recently renovated Baltit Fort, originally built in the twelfth century, in the shadow of the twenty-five-thousand-foot Rakaposhi Mountain. The town has long been the main stopping off point for visitors to the region, but as visitor numbers increased during the 1990s, Karimabad's "character" took a significant turn for the worse, according to many people I spoke to, both tourists and locals. Although there have been concerted efforts to control the incidence and style of new building construction (the Aga Khan Cultural Services is heavily involved and invested in Karimabad's physical and architectural development), there are numerous new hotel buildings whose concrete, often brightly colored exteriors jar with the mud-colored traditional buildings. The main street of Karimabad is now almost entirely populated by those vying for the tourist trade, replete with souvenir shops, hotels, and cafés advertising cappuccinos. During the height of the tourist season, midsummer, the streets are full of people-laden jeeps, beeping their horns as they try to make their way through the narrow thoroughfares, and streams of tourists with rucksacks, clutching their cameras and guidebooks, searching for a hotel and looking for the perfect photo opportunity. Although the women of Hunza do not traditionally practice *purdah*, it is rare to see a local woman in this scene. They keep out of sight in the back alleyways.

I conducted interviews with local shopkeepers and the staff of trekking companies about the various types of tourists they encountered. The backpackers, many of whom are traveling on a "shoestring" budget, are not particularly interesting to local people, indeed, they are often seen as insulting as they haggle for lower prices. Those traveling with local tour-operating companies, often paying U.S.$2,000 for a two-week trip, are more welcome, as they are willing to buy expensive local rugs and other souvenirs, and pay handsome tips. Different nationalities are also recognized as having different characteristics. The east Europeans, I was told, never hire a guide; they are like donkeys, carrying all their luggage on their backs. The Germans are very strict and easily irritated. The English and Americans are the best:

patient, courteous, and generous. The French are cunning, always trying to fleece the guide and the porters.

There are also Pakistani tourists in Hunza, most, like myself, Punjabis, who are easy to distinguish from local people, being darker skinned and usually taller. To local people, these Pakistani tourists are the worst of the lot, rude and disrespectful, always haggling over cheap souvenirs and trying to get glimpses of local and foreign women. The disdain for Pakistani tourists is not simply a result of those tourists' behavior, however, it is also informed by an attitude that relates to the presumed superiority of foreign tourists, both in terms of wealth and taste. Local people in Hunza, particularly those who have extensive interactions with foreign tourists, often emulate those tourists' dress style and general demeanor, working hard to replicate their casual, nonchalant way of carrying themselves and of wearing expensive brand-name outdoor clothes, such as North Face or Patagonia (tourists often leave such items with people as gifts, though they are also increasingly available in secondhand shops in Gilgit). The ideal image is of comfort and style and, crucially, a lack of concern about appearance—though it is, of course, this very "lack of concern" through which they produce themselves as a culturally distinct category of people.

The Pakistani tourists' cheap fake designer clothes, clearly not designed for trekking, and their overt attention to their own appearance, frequently checking their reflection or combing their hair in shop windows or the side mirror of a parked vehicle, like Bollywood stars ready to shoot a song scene, reveal them as the indigenous (particularly "downcountry") face of modernity, distasteful in the eyes of the people of Hunza. It is these kinds of tourists that both the Hunza shopkeepers and the Western tourists want to avoid.

The negative image of Punjabi tourists in Karimabad had an impact on the way my fieldwork progressed. As I have mentioned, anthropologists have visited both Karimabad and Shimshal before, but they have mostly been white Westerners, not Punjabis like me. It did not take long for most local people to see me differently from most Punjabi tourists when they met me, given my notebook, pencil and, perhaps, my "Western" demeanor with the requisite North Face jacket and expensive trekking boots; thus, although my being Pakistani and from "downcountry" obviously had an impact on my work, I did not feel that the Punjabi tourists' stereotype affected my interactions with local people in any significant way. In Shimshal, however, while the Shimshalis themselves had no objection to my spending time in the village asking questions, I did notice that they were reluctant for me to meet foreign tourists in the village. Subsequent conversations with close Shimshali friends confirmed what I suspected—that my presence as a Punjabi in the village could easily be seen by tourists as confirmation of their worst fears about the "wrong kind" of tourist reaching Shimshal. As Shimshalis rely on their untouched status to attract tourists, they work hard to maintain that image even as they welcome interaction and change, as I discuss below.

Barely any foreigners visited Shimshal in the days before the KKH, with only the occasional explorer reaching the village. Before 1974, as we saw in chapter 4, these visitors required the permission and approval of the mir of Hunza before they could visit side valleys like Shimshal. After 1984, a hand-ful of tourists began to visit Shimshal each year, and, in 1986, a hotel, or more like a camping ground with food and tea served, was set up to cater for these tourists by Ahmed Ullah (who in 2006, when I started fieldwork, was running the Ziarat Teahouse), with the ambitious name of Dastghil Sar Inn (ambitious because it really was simply a camping ground, not an inn at all).

The Guest Book: Hospitality, the Road, and the Discourse of Disturbance

My analysis of tourist perceptions of Shimshal begins in the mid–1980s, based on a series of entries in a guest book kept at Ahmed Ullah's hotel. As will be seen, the dominant perception of the village and the valley at this time was as a beautiful, unspoiled paradise of kind and hospitable people. By the late 1980s and into the 1990s, as the road project began to materialize, albeit slowly, and also as the number of tourists visiting central Hunza, though not yet Shimshal, began to rapidly increase, the dominant theme among visiting tourists was their concern about what would happen to Shimshal when it became more accessible to the outside world—including, in particular, the arrival of more tourists.

I was first introduced to the guest book by Ali Hussain, the cook at Hamid Shah's hotel, my residence in Shimshal. One day, I was copying dif-ferent entries made by guests in a new guest book kept at Hamid's hotel, and Ali Hussain asked me if I was interested in something older, and more his-torical; he told me that there was a guest book that had been kept at his uncle's (Ahmed Ullah) camping ground/hotel since 1986. The next day, Ali Hussain brought a ragged and dusty register to me in which were recorded the views of more than one hundred trekkers and visitors to Shimshal since 1986. The first entry was made in 1986, the last in 2003. The entries ranged from a couple of sentences to more than three pages. In most cases the writers gave their name and nationality and the trekking route they did or planned to do. Based on this information, about 88 percent of the tourists were from Europe, about 10 percent from Canada or the United States, and 2 percent from Japan. There was only one entry from a Pakistani.

The first ten or so entries in the book are fairly uniform in their views about Shimshal, focusing on the hospitality of the local people and the serenity and calmness of the valley. As one hotel guest, Steve from San Francisco, wrote on June 20, 1986:

> A one day stop in Shimshal has convinced me that the wider your travel, the better your life becomes! No roads, but this place is bliss. Scenery, hospitality, local food and vibes are excellent. [He continues with a quote from a Bob Marley song—"See the morning sun, on the hillside—if you're not living good, you gotta travel wide."]

More entries from the summer of 1986 follow:

June 20, 1986
I am unfortunate for only being able to spend one day in the beautiful village of Shimshal through lack of time. This valley will have to be revisited when I have more leisure. The climate, hospitality, scenery are all delightful. Mr. G. Baig invited us to his house for local food and in general made our stay a very relaxed one. I am sure anyone who visits Shimshal will receive the same treatment and nobody leaves here disappointed.

July 26, 1986
Unfortunately we only had time for one night stay. The people are so kind and hospitable, we would like to have stayed longer in the house of our porter.

July 28–30, 1986
Great hospitality—Sorry we couldn't stay longer than 1 night before a difficult but rewarding trek to Zardigarbin. Today we return to Malangutti, with a brief stop for some food and of course tea . . . hopefully we will be able to return some day to this delightful valley with its exceptional people.

August 2, 1986
Two penny-pinching students who were narrowly persuaded to hire a guide who brought us up here. We were overwhelmed when a mile before Shimshal people appeared, asking to carry our packs. We were then taken to our guides' house for tea and chapatti—and then to his other house for seconds—heartwarming hospitality.

August 7–9, 1986
Friendly helpful hospitality. It has been a great pleasure staying in Shimshal. I leave with many regrets after meeting so many kind and charming people.

Some of the entries refer to Shimshal as a hidden paradise, a representation similar to that of representations of central Hunza by Western medical professionals in the 1940s through the 1960s in the context of a disease-free and healthy society. Then, as now, the lost paradise status of Shimshal was associated with its physical remoteness and lack of access by motorized transport; the mountains acted as boundaries that kept Shimshalis in and modernization out. Also notice the emphasis on "discovering" paradise. This is, in many ways, a reenactment of the late nineteenth- and mid-twentieth-century official trope of "lifting the veil" of Hunza, which is transformed into the late twentieth-century "trajectory of discovery" (MacCannell 1976, 155) in the global tourism domain. Here Shimshal is represented as the lost world of Shangri-la, a peaceful mountain paradise:

September 29, 1986
God be with them always. Discovering Shimshal is discovering a lost paradise. The people of this valley show sincerity and openness not found in many places in the world. God be with them always.

September 29, 1986
One of the last paradises left in the world for trekkers and adventurers— let's hope they never finish the road from Passu to Shimshal.

November 6, 1986
A short visit, only a day in Shimshal but enough to get a flavor of paradise and its people and mountains. The fall colors are fantastic; reds, yellows; and brilliantly clear days—a place to return and explore more on the mountains. The people have been superb, open, sincerely, hospi-

table, especially as they were not expecting any more trekkers this year. Now back to the road and slowly down the Hunza valley.

August 10, 1987
Finally reached mythical Shimshal yesterday following in the footsteps of our illustrious patron Lord Hunt. Temperament of the people is expressed fully in the previous pages—slightly surprised to find that the sanitation is so basic for such a reasonably large place.

Until 1987, the hospitality of the people, the serenity and calmness of the valley, and the rediscovery of a lost paradise are the main themes of most entries. These perceptions confirm the findings of most social science work on tourism—that people come to places such as Shimshal to escape from their mundane routines and to seek places of untouched nature and history. But as these tourists were escaping their routine structure, they were getting sucked into another structure. This was the structure of newness, as I call it, according to which the motivation to visit somewhere new is perhaps more important than finding an escape. This is apparent in how most of the tourists are "in a rush"—many comment that they are able to stay only for a short period. This, perhaps, reflects Leed's (1991) point that in their efforts to escape their daily routines, tourists often find themselves stuck in another routine, that of the touristic mode, that is marked by an experience of estrangement after estrangement.

During my fieldwork, I asked Shimshali friends to explain the extraordinary hospitality displayed toward tourists who arrived in Shimshal during the 1980s and 1990s. My question made a lot of sense because this tradition was very much on the decline when I spoke to the Shimshalis in 2006, and the people in Shimshal were aware of this as the change was taking place quite quickly and right in front of their eyes. Most of the Shimshalis gave a very simple answer to my question. They said that in the old days when traveling conditions were harsh, Shimshalis could relate to the hardships tourists had experienced during their journey to Shimshal, so they felt naturally compelled to extend their full hospitality to them. This claim is further strengthened by the fact that Shimshal in the 1980s and 1990s was fully self-sufficient in food, so feeding extra mouths had never been a problem, unlike in many other villages in the region where food shortage is a fact of life and hence puts a practical restraint on hospitality. Moreover, hospitality was an extension of an already existing institutionalized practice in Shimshal in which they welcomed "strangers" into their village and homes.[2]

The road project was initially started in 1986, but it was only in the 1990s that it acquired some pace, showing evidence that one day the road would actually reach Shimshal. During the late 1980s and the 1990s, as the road project connecting Shimshal to the KKH began to move ahead, entries in the guest book begin to indicate alarm regarding the impending threat to the local culture from the expected increase in flow of tourists to the village. The tourists become concerned that the Shimshali culture of hospitality, their indigenous values, and indeed the balance between their culture and ecology may be negatively disturbed.

After an initial entry in 1987 that brings up such issues, subsequent entries respond, forming a conversation of sorts, a debate among tourists who likely met one another only in the pages of the book, about the future of Shimshal. Some are more anxious than others that Shimshal should remain "untouched"; others are more pragmatic, arguing that the Shimshalis should be allowed to modernize but that care should be taken in the process (putting the onus for this care not only on the Shimshalis themselves but also on the tourists) not to destroy the uniqueness of Shimshal. This debate among tourists in the pages of the guest book was a precursor to the contemporary conversations and debates that occur among tourists on Internet travel blogs; Shimshal is, indeed, the subject of many such blogs, and the same subject is debated.

The entry that started the debate was made on November 15, 1987, by Alan of Berkeley, California. Alan invokes the discourse of disturbance as he asks what might be destroyed and by what. He lays it out simply: that Shimshal's remoteness, and hence its authenticity and uniqueness, was kept intact because of the very inaccessibility that will end with the arrival of the road, as the friction of distance is decreased.

> As perhaps the last still intact nature culture unity in the region, this town and its people may be more unique and important than all its surrounding magnificent peaks. The absence of the road and the gradual approach to the completion of the road over a decade or so will help to see the inevitable introduction of "modernity" to this nature-culture complex.

Alan is clearly concerned more about the people than the nature, but he does not see the two as totally distinct, either. Rather, he juxtaposes the unique combination of the "nature-culture complex" of Shimshal against the "inevitable" wave of modernity that will disturb and perhaps destroy this complex. What he clearly does not realize is that the arrival of the road does not mean an end to remoteness—rather, it means resignification of that remoteness, from one based on a romantic notion of proximity to nature to one based on certain developmental and sociological aspects, as we saw in chapter 5. Indeed, he represents Shimshal as a microcosm of itself, insulated from the impulses and ethos of "modernity" and marked by a traditional form of social solidarity and organization that has remained unchanged, presumably, because of lack of contact and connectivity to the outside world. He writes:

> Have you noticed the degree of communalism still present here and thought about how that must affect the values and attitudes of the people more generally? It's particularly visible if 10–20 locals drop in on one of the huts while you are there, but evident elsewhere as well. Did you note that while brothers may share a common second name, family names are still generally not used here. Why are there no dogs here and what has become of the locally/traditionally produced footwear (it must have been better)? The rare locally produced thick beige wool trousers are beautiful . . . okay the women's hats too.

Alan also implicates himself in "disturbing" the "native culture" through the introduction of new desires and the ethos of consumerism. He thinks that it is not only the road but also the behavior of the tourists, especially their handing out of consumer goods to the local people, that might change the culture of Shimshal for the worse. While he does want to be generous to the Shimshalis, he also fears that his act of kindness may "disturb" the local society.

> Fine, such giving is consistent with Jesus' teaching and attitudes towards the needy and is quite humanitarian one may say. Granted, but the point is, how can we do this without increasing dependence on these gifts to a destructive or dangerous level. (e.g., traditional skills are lost or abandoned before the means are achieved to buy the replacement goods within one's budget on the "market"). How can we give without reinforcing endless, arbitrary and obnoxious begging? I am afraid I don't have the answers. But if we can't find a better method, I am afraid we won't have to wait for completion of "the road" to see the soiling of this last isolated community.

Alan's lamentation about the "soiling of this last isolated community" reflects a typical reaction from tourists, as we notice throughout the literature on tourism, who see remote and isolated sites losing their charm and attraction as they become physically connected and culturally modernized. According to the tourists, the charm of Shimshal is preserved not only in its geographical remoteness but also in the culture of hospitality, which will be lost with the introduction of new ideas about money, competition, and consumerism. The discourse of disturbance and authentic culture mutually reinforces Shimshal's identity as a remote place in the eyes of the tourists. The fear of change and disturbance is especially sensitive regarding remote areas, as change and disturbance are in some ways the hallmark of city life (de Certeau 1988, 91). The ceaseless flux of the city is the obverse to the ceaseless continuity of the remote.

The response to Alan's entry was overwhelming, the debate he started referred to by almost every future entry. Most of the comments are measured, arguing that perhaps Alan is implicating the tourists and their behavior too much in the potential "disturbance" to the Shimshali culture. The entries also show that some tourists are not oblivious to how the hard living conditions in remote areas affect the health and well-being of the people. Some take a pragmatic position that it is not fair to expect Shimshalis to remain unchanged, while others show more confidence in the Shimshalis' ability to manage the change that will come with more tourists and the road.

September 1, 1988
I am not standing on any pedestal. Shimshal and its people seem good and honest. When the road arrives the place is bound to change, but the people themselves seem to want that. It's not our business to tell them what's good or bad. The hardest three days backpacking I have ever done, I am not an adventurer, 8 students walking to Passu were leaping like Lumnaha, where I had walked one step per minute. This place makes you feel humble.

May 26, 1989
On the whole the transition from the old to the new is being handled better than in most places—but I strongly feel that begging should not be reinforced—these kids are very serious about it and Islam teaches charity to the poor. Most of these people are not by their values poor. A teacher here is trying to record on video to encourage the old culture—it's an exciting idea—may be someone can help him find access to video equipment.

September 25, 1989
It is now three years since I first came to this wonderful place, and this is my third visit. The people are wonderful and the place has the serenity and charm unmatched in the Karakoram (a close second is perhaps Hushe[3] across in Baltistan). The entries in this book up to now show that many people are worried about the effects of contact with the outside world on the Shimshalis. Don't be!! "Progress" will come to this part of the world inevitably though fortunately at a slower pace than Karimabad which suffered a tourist explosion on the opening of the Khunjerab Pass/KKH back in '86. I have sat and listened to concerned individuals going on about tradition being swept away and changes coming for the worst: I agree that coming of the road will perhaps be a sad day—the quiet torn asunder by the revving of a jeep—but it is the people's wish. But "tradition for tradition sake" is a dangerous maxim—lamenting the fact that traditional Hunza/Shimshali houses are no longer built here for example.

Take a look at the soot enclosed ceilings and windows walls and then consider the inside of the people's lungs after a winter spent in such a smoky interior. We came from distant and almost other worldly civilizations to visit this place and are instantly the focus of attention—as wearers of peculiar clothes, potential source of cash, and possessors of extra-ordinary gadgets, strange faces. It should be obvious that the impression the villagers have of us is entirely in our own hands. We are the responsible ones.

These entries reflect that the Shimshalis themselves are part of the debate that is going on in the pages of the guest book and that tourists are discussing these issues with their hosts. Further evidence of this is found in the two entries reproduced below:

July 10, 1990
One should take the time to share with responsible Shimshali people our ideas about tourism and development and our fears for the future of such communities and experience of what has happened in similar areas in other countries (e.g. Nepal). I have found the village leader and some of the people in Shimshal very concerned about the future and responsive to ideas and discussion. Long live a happy Shimshal.

August 13, 1990
After all the difficulties of getting here the people of Shimshal greeted me with their warm welcome. Our excellent guides all have relatives here and we have felt very privileged to be invited into many Shimshali houses to share tea and food with the local people. Talking to many people they very much want a road connection with the outside world and would welcome many more westerners, simply to obtain extra income. It is important that we tread lightly, remain friendly and do not induce the bad habits found on more popular trails throughout the Himalayas/Karakoram. It is never going to be easy trekking to Shimshal, that may be the saving factor that keeps Shimshal one of the last Shangri Las.

The occasional entry expressed indignation at the implication that places like Shimshal would be better left "unmodernized":

August 13, 1992
It was indeed pleasant, rather amusing, to read what the people before have written about Shimshal. With no intention really to criticize the perverted and exploiting kind of us all who come here under the guise of tourist, I would like to remind everyone that I have got no business to choose what is best for them. I feel everybody who comes here must have that attitude. I would like Mr. Alan Blair [the trekker who wrote the comment of November 15, 1987, that started the conversation about the pros and cons of a road connection to Shimshal] to live here just for one year with his wife, children, old parents and brothers and sisters, cut off from the rest of the world without any medicine when one of his dear ones is dying and so on and on. . . . He can only then answer why a child ask[s] for a pen or a pantaloon. He can only then feel whether a road is more important or whether he would prefer Shimshal being stored here as a guinea pig for a case study by the rich and spoilt.
Mr.—KEEP YOUR COMMENTS WITH YOU OR CARRY THEM BACK TO PASS ON TO YOUR SONS BUT LEAVE SHIMSHAL ALONE. LET THEM CHOOSE FOR THEMSELVES.

A similar tone in the entries continues in the 1990s. The road continues to be a topic of discussion as well as the lack of electricity, in some ways the ultimate hallmark of modernization, with tourists at once recognizing the importance of these facilities to the Shimshalis but expressing regret at what will be lost:

July 20, 1997
I have had a wonderful stay in this beautiful village—such a contrast to
the dryness and desolation of the road—but the slog was worth it. The
hospitality of the village has been extraordinary. I am glad I made it
here before the road and only 2 weeks before electricity! I hear there is
going to be a big celebration.

In general the emphases on hospitality and the threat to local culture
remain central in tourists' comments into the late 1990s:

July 4, 1996
Till now I was especially impressed by the lovely place name Kuk opposite the Malungutti glacier. It was a little paradise, sitting in a natural
bath in the middle of the blooming bushes and trees looking out of the
magnificent tops of Distgil Sar (7885 m) and nearby peaks, also the
thundering sounds and sights of the glacier breaking down into the river
was "thrilling." Just because of Kuk (and then I didn't say anything of
other special moments, meeting with friendly people, the huge rocks, the
rivers etc.) I hope to come back to this wonderful valley (may be by jeep?
I hope it for the Shimshal people!).

September 1998
En route down to Passu having spent a wonderful ten days up at the
Shimshal Pamir. Climbed Zoiduri Sar (5,500 m) and Zarsenic Sar
(5,900 m). . . . Very much enjoyed our base camp besides the two lakes
and have also enjoyed the welcome and hospitality shown to us by all
the Shimshali People in Shimshal and Shewert.

The entries then return to the theme of how best to minimize disturbance to local culture. Here again, the Western tourists prefer their own
advice as they have already seen and experienced the modernity that
Shimshal is about to experience through the opening of the road.

June 1999
As for progress and Shimshal, it's now ten years on from some of the
comments written earlier in this book and the road still isn't here. Will
it ever be, given those huge ever-shifting scree slopes? However electricity has arrived—a year ago—and one house even has satellite TV.
I think it's desperately difficult for an outsider to gauge the real effects
of new technology on a particular culture (read the debates in our press

on the effects of TV on our Culture—let's face it, we are not too sure how technology changes us sometimes, are we? Still, we have a wider experience of these matters than the Shimshalis, via travel, education, etc, so I think it is OK for us to voice any concern we may have. Remember, a choice is only a choice if it is informed. Also remember that development is about creating a happy and sustainable community for its inhabitants—not a beautifully arranged fishbowl. May be the biggest threat to the society here is growing disparity of wealth. Residents of Britain and America could be the best people to warn Shimshalis of this! Anyhow, enough theorizing and speculating. Shimshal is lovely, its people are amongst the nicest anywhere, and the scenery is fantastic. Thanks to Yahya Baig, Gul Mohammad, Mohammad Jan and everyone else for showing me *their homes, their food and their culture* [emphasis added] See you in '00!

Other tourists argue that the arrival of the road is inevitable and one can only hope that this brings prosperity and happiness to the local people:

July 2003
We—Medy and Cees Raven from Holland—arrived at Shimshal a few hours ago and already we met many friendly and helpful people. Tomorrow we will look around and rest. The beginning of road from Passu to Shimshal was amazing and in much better condition than KKH! You can call it "Shimshal Highway." We hope that soon all bridges will be ready and that the direct connection will bring you much prosperity.

The entries reveal some ambivalence among the tourists regarding how increased connection with the outside world might change Shimshal. Some show confidence that Shimshalis are well prepared for the imminent change, while others are not so sure. Posing the question of the road in this fashion shows the tourists' implicit assumption that because Shimshalis have remained unchanged and undisturbed throughout history, they may have little ability to cope with change or "disturbance." Hence, many tourists think it is the responsibility of the outside world, the world of flux, to help them. To some extent the tourists' concerns, if not their assumptions, are justified, as the arrival of the road represents a rupture in Shimshal's history. But Shimshalis have experienced changes of other kinds and other magnitudes in their lives; indeed, their very encounter with the tourists who are writing about the dangers of such an encounter to the Shimshalis is a testimony to their ability to adapt to change. The invasion of the British, the retrieval of Kashmir forces at partition and incorporation into Pakistan, the invasion and temporary occupation of the Shimshal Pass by China— all are events that shaped and changed Shimshal and to which Shimshalis responded. But in the discourse of authenticity and disturbance in the late twentieth century, outside tourists didn't recognize this history of Shimshal, and even if they did, it did not seem to have an effect on their perception of Shimshal as a remote place, insulated from change. Their approval of letting Shimshal decide for itself what level of modernity to embrace was based mostly on abstract principles, such as freedom to choose and self-determination, rather than on the historical experience of the Shimshalis. How the road did change the character of Shimshal and whether the tourists' fears and predictions materialized will be discussed in the next section.

The Tourist Industry and the Changing Nature of Hospitality

Before we analyze how the Shimshalis responded to the arrival of the road and how it changed the character of Shimshal's remoteness and culture of hospitality, I would like to make one small point. Most anthropological studies of tourism look at how tourists' expectations and choices of destination are shaped by the tourism industry's marketing strategies disseminated through brochures, Web sites, television and newspaper ads, travel channels, and guidebooks (Molina and Esteban 2006; Andereck, 2005). The case I present here is slightly different because most of the tourists who visited Shimshal between the 1980s and early 2000s were not influenced by the national or international tourism industry. Most of them came to Shimshal after hearing about it from local people in Hunza. Thus, the image of Shimshal tourists express in their entries in the guest book was not influenced by expectations mediated by the tourism industry's representation of Shimshal. Rather, their perception was created by the tourists themselves, and perhaps it would be correct to say that these representations of Shimshal were made for the consumption of both the Shimshalis and fellow travelers. Later on, during the early years of the 2000s, we see that these representations were picked up by the tourist industry in Pakistan and elsewhere—a reversal of the usual process, in which the industry shapes the motivations and destination choices of tourists.

If one Googled "Shimshal" and "trekking" in 2007, one got more than three thousand hits.[4] In recent years, the Internet has become a key site for people to share ideas about such places, both through Web sites advertising trekking holidays and through blogs in which tourists relate their experiences. A Pakistan-based trekking company run by a man from Skardu has this information on its Web site about Shimshal: "This last remnant of true *Hunza culture* exists near Pakistan's Pamir east of the town of Passu in upper Hunza valley. . . . The reason why Shimshal was located in such an out-of-the-way place is because this is where the Mir (ruler) of Hunza used to send his people who were found guilty of crimes and also those who he just did not like! Shimshalis now are a proud and hardy folk. Their *hospitality is legendary* and the natural beauty that they inhabit makes us feel that perhaps the Mir sentenced them to heaven, not hell!" (Concordia Expeditions n.d.; emphasis added). A Nepal-based company that markets treks through local tour operators in Pakistan has this to say about Shimshal: "Shimshal is the oldest of the Wakhi-speaking villages in Gojal. Established over 400 years ago, the 1200 residents maintain customs no longer practiced in other villages. Shimshal was closed to outsiders until 1986. Since then, only handfuls of foreigners have made the trek each year" (Trek to High Mountains n.d.). Another local company, run by "youngs" from Hunza, writes: "It has attraction for the society lovers as the *only place where the culture and values of Hunza is still alive.* These all make Shimshal as one of the best place to visit in Hunza" (Karakoram Adventure Holidays n.d.; emphasis added).

These Web sites represent Shimshal in terms strikingly similar to the entries in the guest book, with the emphasis on hospitality and the preservation of authentic "Hunza" culture[5] against the disturbance of modernization. In some ways, these representations are about a decade late, in the sense that by 2008, when these passages were recorded, Shimshal had already undergone drastic changes due to the opening of the road. Notice that nowhere do the tourist companies mention that now there is a road linking Shimshal with the outside world, although they do mention that since 1986 only a handful of people have visited it. They do, however, emphasize the unique status of Shimshal as the "last remnant" of a dying culture, a theme very similar to the discussion in the guest book. The companies' use of these discourses of a dying culture is paradoxical, as by encouraging consumption of these places, they are encouraging the process by which such places will no longer exist.

Tourists' concerns about disturbance to Shimshal's culture and the tourist industry's continuation of this discourse are not, of course, invisible to the Shimshalis. Just as the people of Hunza have tried, and continue to try, to meet the expectations and fulfill the desires of the tourists who come to Karimabad, so the people of Shimshal do the same thing; though the example of Karimabad reminds them in many respects what *not* to do. Methodologically, one of the effects of this was the Shimshalis' reluctance to let me interview foreign tourists. Perhaps my presence in Shimshal, the Shimshalis thought, would undermine its image as a remote, serene, and idyllic village, an image that both the Shimshalis and the tourists wanted to maintain. The desire to maintain this perception was intensified after the completion of the road in 2004 and the dying of the tradition of hospitality of inviting foreign tourists into their homes. By 2009, tourism in Shimshal had been fully "commercialized" to cater for the large number of tourists visiting Shimshal. The tourists were still attracted by Shimshal's remoteness but not its hospitality. With the arrival of the road, tourism took a slightly different turn: more and more adventure tourists, attracted by Shimshal's mountain scenery and trekking opportunities, started coming in, as opposed to mostly cultural tourists. The Shimshalis seem to welcome this shift.

The opening of the road and the arrival of tourism in Shimshal also affected the transhumance migration and cycle. As I write these lines, the tradition has changed drastically: in 2011 only ten households moved to the Pamer. Yaks, which earlier carried rations for the summer and winter herders, now carry the North Face rucksacks of foreign tourists.

In some ways the discourse of disturbance that the tourists expounded in the pages of the guest book did materialize. Competition over who will go with the trekkers as porters, whose yaks will be used for carrying trekking equipment to the Pamer, whose jeeps will be used for transportation, and in whose accommodation the tourists will stay has become commonplace. It seems that the earlier hospitality of the 1980s and 1990s, which was informed by the Shimshalis' own experience of the hardships of travel, has been replaced by hospitality informed by the standards and norms of the tourist industry. It is not that Shimshalis have become less polite or friendly toward foreign tourists. In fact, they may have become even more polite and hospitable than before. Rather, their behavior is informed by a different cultural logic, a logic that the earlier tourists were trying to escape.

This is the logic of the tourist industry, of capitalism, in which human sentiments, desires, and emotions can all be turned into commodities. The earlier human relationship of hospitality between tourists and Shimshalis, although by no means purely noncommercial, was based on an extension of an important cultural tradition in Shimshal; now it is being transformed into a mostly economic relationship of hospitality, informed by the norms and ethics of the tourist industry. True, the earlier form of hospitality was also a part of a capitalist economy, but in that case, the Shimshalis' transhumance cycle of migration and its attendant customs, such as taking care of travelers who come from a long distance, were fully practiced, so it was not difficult for the Shimshalis to adapt this custom to the newly arriving tourist class. With the arrival of the road, the transhumance economy is in decline and new economic professions are becoming popular. Hospitality that emerges within this situation does so in the form of a commodity whose production and exchange engender a new kind of relationship between tourists and Shimshalis. From the perspective of the Shimshalis, at least, part of the relationship enshrined in the earlier practice of hospitality was about tired strangers in a distant land looking for the comfort of a home—more or less a subjective moral relationship. The newer concept of hospitality, however, is an objectified economic relationship in which hospitality is produced by the Shimshalis and consumed by the tourists. Is this a good thing or a bad thing? Is Shimshal's slide toward commercial tourism sustainable? These are difficult questions whose answers will be another decade coming, based on evaluation by the Shimshalis. It will perhaps depend on their ability to balance the commercial economy of Shimshal with the subsistence-oriented economy.

I will now present a case of how commercial tourism is transforming a traditional Shimshali house.

Constructing Hospitality: The New "Traditional" House

One summer evening, Hamid Shah invited me to his house for dinner. Hamid's house, adjacent to the hotel compound, was a traditional Wakhi house surrounded by fields and fruit trees. The outer boundary of the house was a square shape, with its internal space divided between sleeping areas, an eating area, a central hearth, and a storage area. The traditional Shimshali house is a complex and complicated internal space symbolic of clear social boundaries and hierarchies. Where one sits is determined by one's social position with respect to the rest of the persons in the room. The five different divisions within the rooms, marked by the height of the floor

Figure 12: A Wakhi house

and the placement of five pillars, reflect the status and culturally sanctioned respect of the people who sit there. Such internal division and control of space is a symbolic representation of the communal and cosmological order and hierarchy of the Shimshali universe.[6]

In terms of its basic layout and function, the traditional Wakhi house combines all residential activities, social and individual, in one open but connected space (see figure 12). The central hearth is perhaps the most revered place in the whole house as it symbolizes sustenance and continuity of the household. At the entrance of the house, just where one enters the door, is a little space, *kinj*. After passing through this, one enters the main open-floor part of the house, called *yorc*, which is the lowest level as well as the place where all footwear has to be left. This central area is surrounded on three sides by elevated platforms that are used for sitting during the day and for sleeping at night, the *nisine raz*, *kla raz*, and *pust raz*. These platforms are built of mud and are covered by straw and rugs woven from yak and goat hair. During the day, the beds are rolled up and stored next to the retaining walls. On the fourth side of the yorc, toward the central hearth, is an area slightly raised from the yorc but about five centimeters lower than the surrounding raz. This is the *nikard*. In front of the nikard is the *dildung*, a platform raised about five centimeters higher than the raz (so the highest platform in the house) where the cooking is done. Usually there is also a slightly elevated platform, below the dildung but higher than the nikard, called the *dildung-e-ben*, where informal guests are served food (this is usually where I would be asked to sit when invited into a house).

According to the Shimshalis, all these spaces are differentiated by variables of function, gender, and degrees of honor, and thus the internal space is inscribed with power and power difference. This aspect of the house is not obvious in the everyday life of the Shimshalis; it becomes most manifest on occasions of social, religious, and agricultural festivals and ceremonies. On such occasions Shimshalis keep a keen eye on the seating order and location of their guests, as negligence to assign an appropriate place to a person could result in strain or breakup in social ties. In general, those with political status are seated on the right side of the central hearth on the raised platform, the nisine raz, which offers the maximum of comfort and a good view, while those with religious status are seated in the kla raz, from where they can see people entering the house. In public and social events such as wedding ceremonies, the nisine raz is filled with the political and religious leadership of the village, the nonleadership members of the visiting family are seated in the kla raz, and all the members of the host family are in the pust raz. The space behind the cooking platform (dildung) is reserved for women to work. In an extended household system, with two or three married sons and their families living in the house, the space is further divided according to the normal patriarchal social hierarchy. The head couple of the household would sleep in the nisine raz and the eldest son in the kla raz. A newlywed couple would get a place in the pust raz at the back of the house. The children would usually sleep with their parents or grandparents.

The central hearth, *buxari*, is located under the *kut*, also known as the *paanj tunni* (*paanj* means five and *tunn* means being or bodies), an opening in the roof with five overlaid pieces of wood. This acts as a chimney for the smoke rising from the hearth (wood is used as fuel). The paanj tunni represents the divine family of five: Muhammad; his daughter, Fatima; her husband, Ali; and their two sons, Hassan and Hussain. These are the ancestors of the Ismaili (and Twelver Shia) imams. Placing the central hearth under the paanj tunni links the productive and sustenance-providing part of the house to the divine family of five for which, according to Ismaili belief, the entire universe was created. Thus the internal structure of the house is actually connected with transcendental authority. To the left of the hearth, and within the main residential quarters of the house extension of the kla raz, is the *zica raz*, where a new mother moves after having a baby for nursing and sleeping. When not needed for this purpose, this area is used for storage.

There are a total of seven pillars in the house, each of which has its own religious and symbolic significance (see also Steinberg 2006, 140). The pillar to the right of the hearth in the nisine raz represents the political (Muhammad), while the one on the left in the kla raz represents the spiritual (Ali) leadership of Ismaili Muslims. In public and communal events, one would notice the significance of these pillars coming to life. At weddings, funerals, or festivals, the political leader of the village would sit by the right pillar, whereas, the religious leader, the *mukhi*, would sit by the left pillar, representing the symmetry between conceptual and actual representations of their significance. An earthen oil lamp is kept in a niche in the left pillar, which symbolizes light as a metaphor for spirituality or inner light, *noor*, associated with the current imam himself.

The pillar on the right side of the hearth is where the first harvested sheaf of barley is hung, marking the continuity between the field and the home. Where the two pillars stand at the entrance from the kinj into the yorc, flour is sprinkled on the first harvest day. There is also a hierarchy and order according to which certain sprinkling rituals are performed. For example, on Kit-e-datt, a festival celebrated on February 15 and generally associated with spring cleaning, the oldest of the skuewene fills the house with smoke of a wild bush, spunder, to chase away the winter. After this he sprinkles flour on the paanj tanni, followed by the central beam, then the pillars, and finally the inner walls.

One morning in October 2005, I noticed that Hamid Shah had started digging foundations for a new building in the western corner of the land where his hotel was built. He was building a traditional house, he told me. "But you already have a traditional house," I said to him. "Yes, but this one is for the tourists," Hamid replied. Over the next few days, Hamid and his friends debated the design and internal layout of the house, involving me in the discussions, which often took place outside and sometimes inside my room. The debate focused on how best to design the house so that it catered to the needs of the foreigners but at the same time gave them an authentic experience of Shimshali home life and culture. Hamid and his friends were aware that tourists were keen to experience local culture; they had all met tourists eager to enter their homes. In addition to this, the tourism industry in Hunza has in recent years taken a "cultural" turn, with the multimillion-dollar renovation of the central Baltit Fort, now the most popular destination for tourists, and strict controls on changes that residents may make to their homes so that the traditional look of the village will not be spoiled. The question for Hamid and his friends was how far they should modify the internal design of the "new traditional house"—too little, and the tourists might not want to stay there, too much, and they might find the experience inauthentic.

I present here an account of each design modification in the house that was debated, how the issue was settled, and what criteria were used to justify the modification. I will present an analysis of the debate and the final design, highlighting the main structural differences between the old and new traditional house, which also reflected a social structural change.

The first thing that was debated was how the house should be used. Hamid was adamant that it should be used for sleeping as well as dining and general relaxation; if the intention was to provide foreigners with a true experience of the Shimshali home, then they should be allowed to use it in the same manner that Shimshalis use theirs. The lack of private sleeping quarters in the house was a key problem to be addressed. Hamid's cousin and the cook, Ali Hussain, maintained that the traditional house should be available not just to one group exclusively; unrelated groups of tourists might share the same roof. Ali Hussain pointed out there might be couples in those groups and those couples might want to have sex. "It's okay for us Shimshalis to stay like that," he laughed. "We have the quietest sex in the world. But if these foreigners have sex here they will wake up everyone else in the house!" In a "true" traditional house, having sex would not have been a problem. In this house, however, inhabited as it would be by groups of foreigners, it became an issue both because of the perceived inability of those foreigners to restrain themselves (have sex quietly) and because some of them would be strangers to each other, which would be very unusual in a traditional house. The toilet, too, was a key problem. In traditional Shimshali houses, there are no toilets; the adjacent cowshed or the fields are used for that purpose, as in many rural places all over the world. Hamid and his friends were well aware that this arrangement would not do for tourists. So it was decided that a separate toilet would be constructed outside the new traditional house.

Another thing that was debated was the location of the central hearth and whether cooking should be allowed within the main living quarters, as is done in an actual Shimshali house. Again, Hamid Shah wanted to provide the tourists with an authentic experience, arguing that they should be able to cook as the Shimshalis cook and have the authentic experience of preparing meals in a real fireplace. Hamid's brother-in-law and the manager for the hotel, however, disagreed, proposing that cooking should be done away from the places where people sit and eat. "Our spices are too smelly," he said. "Anyone sitting in an area where our food is cooked smells of turmeric and cumin for days." He also pointed out that Shimshali cooking methods might be considered unhygienic by tourists from Western countries, who the Shimshalis see as obsessed with cleanliness when it comes to food. Hamid's wife suggested that perhaps they, the hosts, could prepare food in their real traditional house and then serve it in the new traditional house for the guests, but Hamid and the others decided that that would be too time consuming and inefficient, particularly if guests wanted tea, coffee, or other small items instantly. So a decision was made to build a separate room called a kitchen at the back of the traditional house, connected to the main room of the house with a window opening through which to serve food. A kitchen is not part of a real traditional house; in fact, Hamid told me there is no word for "kitchen" in the Wakhi language.

Then the internal division of the house came under discussion, especially the height of the sections of the house. It was unanimously agreed that there was no point in having five different heights marking seven different sections of the internal quarters of the house as in their own homes. Hamid worried that these height differences would create a nuisance for the tourists, perhaps leading to accidents, as people not used to such spaces might trip and fall. But more important, all agreed that the purpose of the height difference in the traditional house, to mark the hierarchical social order within different social categories of the community, would be meaningless to foreign tourists. This house would have only two social categories, guests and hosts. (Although Hamid and his staff or relatives would not be staying in the house with the tourists, they did expect to spend time there, not only serving them but also chatting, drinking, and dancing with them.)

All the discussants agreed on the need to include a paanj tunni, both to provide sunlight and because of its religious significance. Although the new traditional house was to be a space for conducting commercial activity, the discussants agreed that it still needed the paanj tunni—to bless the business.

When I returned to Shimshal in the spring of 2006, Hamid had finished building his "new traditional house." Based on their discussions as outlined above, the major differences between this house and a real one was the lack of the elaborate system of varying heights between different portions of the house (the new house has only two different levels) and the separation of the cooking area, now called a "kitchen," from the living quarters. There were no separate sleeping quarters and, as yet, there was no separate toilet. For the time being visitors were expected to use the toilets in the main hotel, but Hamid had plans to construct a toilet in the house in the near future. So far no one had stayed in the house, but Hamid was keenly looking forward to the summer, when he expected it would be popular among tourists.

The issues about how far the new traditional house should cater to the "modern" needs of tourists reflect Shimshalis' wider concerns about how their material and social lives may be perceived by the outside world, now only a jeep drive away. Indeed, the issues also reflect how they themselves are beginning to see the way they live in a new light, questioning whether and how they should adapt accordingly. The questions raised in the construction of the new house relating to privacy and hygiene are characteristics associated with a more modern form of social existence, associated with a Western lifestyle, in which bodily activities are slowly pushed from the public and more shared space to private spaces (Elias 2000). In order to deal with this sensibility, the new house either had to be an entirely public space, with cooking and bathroom needs shifted to external spaces when used by mixed groups, or an entirely private space when used by a couple. Although Hamid's new traditional house most starkly represents these changes, as it was built specifically to cater to people for whom that "modern" way of life is considered normal, many Shimshalis are slowly making changes in their own homes, with the construction of separate rooms for married couples, separate kitchens, and "Western" toilets. It is interesting that in their discussions about how to design the new traditional house, Hamid and his friends did not refer to these changes taking place in houses throughout the village—the "traditional" house they had in mind was the unchanged, ideal version (see Adams 2005).

Shimshalis' reduced dependence on the "natural" subsistence economy is, I suggest, reflected in the changing position of the kitchen. In the "real" traditional house, the hearth is a symbol of household subsistence that links society and nature; on the central pillar of the hearth, new crops, new butter, and new flour are sprinkled to mark the connection and continuity between nature and society. In the new traditional house, the hearth matters less, just as subsistence and the link between nature and society matter less, the link between a market economy and society replacing the older nature-society relations. Of course, this is not to say that earlier Shimshalis did not participate in the market economy, as the flow of tourism even in the 1980s testifies to this.

If the traditional house is about continuity, connections, and hierarchy, the new house is about exchange, divisions, and different kinds of hierarchy. In the traditional house, connections are maintained and sustained through hierarchy; the different levels of the floor represent different positions in the social order, which in turn represent a hierarchy uniting the people in the symbolic space of the house. The new traditional house does away with the different levels as there is no need for this space to depict the social order of the village. The new house and its levels are based on different criteria, those that cater to bodily activities such as sitting and eating. The erasure of floor levels in the new house facilitates the flow of exchange between tourists and Shimshali entrepreneurs. Hamid expects tourists to use the raised level for sitting and drinking or chatting, while on the lower floor there will be a dining table for eating. This is traditionally the place of the women, but their role is removed from the new relationship: women, who represent house-holds' traditional hospitality, are not needed in the new space because this is the hospitality of the hospitality industry.

Conclusion

As Hunza became more accessible with the building of the KKH, the social geography of the region changed. Villages and towns of Hunza located on the KKH or within easy traveling distance of it became connected to the tourism industry and the infrastructure of state and development organizations. The isolation of Hunza behind mountain barriers, which had been a major part of its appeal to earlier visitors, such as colonial explorers and frontier officers, medical professionals and farmers, was now gone. In the process, these parts of Hunza lost their appeal to the modern-day equivalent of such visitors, who by the 1980s were mostly simply tourists. In this context, villages such as Shimshal, which remained not easily accessible, still stuck behind mountain barriers, then acquired a new significance for tourists and other visitors. But a similar process ensued in Shimshal, too.

Today, Shimshal's internal space of separation and isolation, informed by the transhumance economy, is being reconceptualized by the Shimshalis as a territory of economic opportunity. The remoteness of earlier times, tied to the culture of hospitality, is slowly changing to a different kind of remoteness and a different kind of hospitality. Remoteness is now conceptualized as a touristic quality, and hospitality is being thought of in terms of the "hospitality" industry. The marketing of new trekking routes has objectified Shimshal's remoteness, codified it in trekking stages and itineraries. Global tourist itineraries are being superimposed on the indigenous transhumance itinerary. The circuits of circulation of yak and tourist have overlapped. Remoteness is now being constructed specifically for busting it, for overcoming it. The remoteness of separation and isolation is replaced by the remoteness of plans, timetables, and connections. But the adherence to strict timetables is structured by a different logic, that of capitalism. No longer is observance of schedules a practice for decreasing anxiety; rather, it is done for increasing efficiency—and profit. Each stage is a number of rupees.

By the end of the first decade of the twenty-first century, Shimshal village had also become a "disturbed" place; it was no longer the final destination for many tourists, rather, it became the starting point for other "remote" destinations, leading tourists to the vast internal space, the summer pastures, of Shimshal. This has pushed the tourists deeper into the heart of the Karakorums, the domain of wildlife. Indeed, this is how Shimshal is marketed today—by the promise that if you go further into the remoteness, you reach the realm of the wildlife, of nature. Here is an excerpt from a Pakistan-based Web site: "The trek from Shimshal to Kuksel takes us into a very remote area and gives us a chance to see some of the endangered wildlife in the Khunjerab National Park. If we are lucky we may be able to observe Markhor, Ibex, Marco polo, Sheep and even the elusive magnificent Snow Leopard" (Karakoram Adventure Holidays n.d.). In the last chapter of the book, I explore how Shimshal's remoteness is constructed in the discourse of global biodiversity and conservation of wildlife.

CHAPTER 8

Romanticism, Environmentalism, and Articulation of an Ecological Identity

Since 1975, Shimshal has been embroiled in a conflict with the Gilgit-Baltistan Forest Department over the establishment of the Khunjerab National Park (KNP) on its traditional grazing grounds. The KNP was established on the recommendation of the famous American naturalist George Schaller, who in the 1970s visited the upper Hunza valleys and saw the land-use practices of the local people as a threat to Himalayan wildlife and nature. Schaller was particularly interested in the plight of the fabled Marco Polo sheep, which he thought had come under threat from local herders who graze their livestock in the windswept valleys of the Pamir and Karakorum Mountains in upper Hunza (Schaller 1988, 98). The KNP remained on paper only until the late 1980s, at which time the government, with the help of international conservation NGOs, mainly the Worldwide Fund for Nature (WWF) and the International Union for Conservation of Nature (IUCN), tried to enforce park regulations, which included a total ban on local communities using the area for grazing. Eight villages were affected by the KNP regulations and all of them resisted any enforcement.

In 1995, after persistent stiff local resistance, the government drew up a management plan for the KNP in collaboration with the affected communities in which it recognized some of their rights, such as limited grazing and fodder collection, sharing of the park entrance fee, and priority to local people in employment in park administration. A consensus was reached among seven of the eight communities, which agreed to work with the government under the revised park regulations. The village of Shimshal, however, refused to cooperate or give up any land to park management. The Shimshalis argued that they suffered more than any other community from the establishment of the KNP, as their lost land constituted more than 90 percent of their traditional grazing area. To date, they remain at logger-heads with the government over the park. One of the ways in which Shimshalis have engaged in the conflict is through remaking the issue from one of control over environmental resources into one of control over cultural identity. They argue that their cultural identity is tied to their liveli-hood practices, which are in turn tied to their access to grazing areas—thus, losing their land to the national park amounts to losing their unique cultural identity.

In the final chapter of this book, I examine this tension between the threat of the KNP to the cultural identity of the Shimshalis, as they repre-sent it, and the threat of the Shimshalis to wildlife and nature, as repre-sented by supporters of the park. Using environmental conservation as a socio-spatial domain, I show that Shimshal is constructed as a remote area and society within two structurally different discourses of nature conserva-tion that often have very different material consequences for a remote and marginal community like Shimshal, where people depend on "wild" natural resources for their livelihoods. Although in both discourses, Shimshal's geographical remoteness is constructed as remoteness from culture and hence as proximity to nature, there is a difference in how the Shimshali people are seen: either as separate from nature or as part of it and hence as threatening to nature or not. Based on this difference, the two discourses often foster competing versions of conservation policy and practice.

First, I will examine the discourse of nature conservation that calls for a complete removal of humans from a conservation area. Nature conservation in this discourse is to be carried out by literally removing an area or resource from human use, setting it aside for wildlife and the aesthetic enjoyment of people who are primarily from nonremote areas.[1] Remoteness in this discourse, then, emerges as a category of refuge—refuge for nature from society's inexorable consumptive demands (though the consumption of the aesthetic is welcomed). I show that once conservation areas are identified, institutional and structural forces exacerbate their condition of remoteness.

Second, I will examine the discourse of nature conservation in which communities like Shimshal are imagined as part of a nature-culture com-plex. This discourse represents such communities as sufficiently removed from mainstream society to be unaffected by its norms and ethos, part of nature and in sync with its processes and hence not a threat to it. I will focus, in particular, on how the Shimshalis have responded to and strategically used this second discourse of nature conservation to fight the dispossession of their grazing lands that resulted from their subjection to the first type of discourse. I show how during the 1990s and 2000s the Shimshalis have fought the KNP regulations by using their cultural identity as a remote and unique society whose livelihood strategies have remained unaffected by the wider forces of the market economy and globalization.

Nature in Remote Areas

To modern society, a remote natural setting is a space where nature in its sublime form can be experienced. Modern conservation is based on the explicit assumption that nature and wilderness cannot coexist with humans, so conditions of complete wilderness have to be created by keeping humans out. Brockington, Duffy, and Igoe state, "The pursuit of wilderness promotes an ethic in which the only landscapes worth saving are those that are distant and exotic, while landscapes that are proximate and mundane appear unworthy of our concern" (2010, 9). Commenting on the association between remoteness and conservation, Neumann writes, "National Parks represent remnants of pre-European landscapes, pockets of remote, unoccupied wildlands" (2002, 28). Deliberate policy decisions to make a conservation area as difficult as possible for human use actually create a condition of geographical remoteness. When conservation units are established, accessibility to them is severely curtailed and controlled, so remoteness is constructed and perpetuated rather than just being a preexisting condition of an area. This is what Roderick Neumann has called "imposing wilderness": the creation of wilderness by political intention rather than natural circumstances. Conservation practices in national parks create remoteness in the classical sense of the word, that is, by *removing* areas from daily human use.[2]

The idea of wilderness lies at the heart of the modern nature conservation industry and discourse. The emergence of the concept, as we understand it today, is fundamental to the processes of globalization and capitalist expansion. Remoteness, as we have seen throughout the book, is a category that has been dialectically engaged with exploration and expansionism. Thus, just as the idea of remoteness is constructed in the very processes of exploration and expansion, so the idea of wilderness is constructed in the processes of industrialization and globalization. The expansionary and exploratory impulses of modernity and globalization are structured differently, however. Remoteness is produced before the territory on which expansion is to be carried out has been exhausted, whereas wilderness is constructed after it is. Max Oelschlaeger writes that wilderness comes on the heels of the "total humanizing of earth's landscape." He further notes, with reference to the United States, "As the nineteenth century began, wilderness existed in abundance; even on the East Coast, great expanses of land, timber, and water were yet to be encroached upon. By 1900 demand for wilderness was beginning to outstrip supply" (1991, 3). As shown by Roderick Nash (1967), Frederick Jackson Turner's frontier thesis made it clear that wilderness was central to U.S. identity, so as it became scarce, efforts ensued to maintain and preserve it.[3] The importance of wilderness to U.S. cultural identity, society, and economy can be linked to the creation of the national park, in which large tracts of land were set aside from human use and destruction. The idea of national parks was driven by two complementary but different factors: reserving land from economic expansion and celebrating wilderness. Social scientists largely agree that often wilderness is socially constructed in the very process of its protection. Roderick Neumann has introduced the concept of the "national park ideal": "'nature' can be 'preserved' from the effects of human agency by legislatively creating a bounded space for it controlled by a centralized bureaucratic authority. This model was initially implemented in the nineteenth century United States at Yellowstone and has subsequently been disseminated worldwide . . . I also use the national park ideal to refer to a particular conceptualization of nature that has been prevalent in the Yellowstone model since its inception. It is a conceptualization of nature that is largely visual, that treats nature as 'scenery' upon which aesthetic judgments can be laid" (2002, 9).

The national park ideal differs from ideas of wilderness in the early nineteenth century, when nature was appreciated for its resources. Earlier, the English concept of the pastoral came to signify a particularly nature-based aesthetic, but that aesthetic did not promote nature as something removed from society—rather, society was part of nature in a seamless way (Worster 2006; Oelschlaeger 1991). Conversely, nature standing apart from society became the core of the national park ideal.[4] With emphasis on the absence of humans, wilderness is primarily valued for its scenic beauty and appeal to certain aesthetics.

The establishment of national parks has conventionally been driven by the aesthetic idea of nature as sublime. Particular landscapes are viewed as awe inspiring because of their vastness and grandeur—and the absence of humans (Cronan 1995; Neumann 2002). In this view, nature is pure, pristine, and untouched by humans. Naturally, then, this discourse of nature conservation maintains that ideal candidates for the establishment of national parks are certain remote "places that remained as they had been before humankind evolved" (Dowie 2009, 249).

While it is true that the initial national park ideal in the United States was primarily a landscape-based aesthetic in which land was seen as scenery, later on we notice that wilderness and national parks come to be associated more with wildness and wildlife. This shift was most pronounced in the early twentieth century when the idea of national parks made its way to Africa, where colonial hunting and land-use practices and policies had led to a decline in wildlife. "First, the parks were intended to play a role as sanctuaries for the declining wildlife population. Human population growth and over-hunting were cited as the major threats, and it was argued that only parks and reserves could save some species from extinction" (Neumann 2002, 124).

Between the establishment of Yellowstone and Serengeti the idea of wilderness underwent a transformation from one based on an empty landscape to a landscape full of wild animals. Note that both the buffalo and the wolf had been eliminated from Yellowstone when the park was formed. It was not until the second half of the twentieth century that these animals were reintroduced. So the initial impetus was to protect the landscape, not the wildlife.[5] Establishing national parks to conserve wildlife increased the reach of conservation institutions and the state into remote pockets of natural wildlife refugia. Remote areas with low human density provide more habitat for wildlife, especially those areas where the difficult terrain makes hunting less prevalent. In some ways these areas and their people are penalized for protecting wildlife that is threatened elsewhere.[6] This discussion sufficiently sets the stage for understanding the history of contemporary wildlife conservation in Shimshal. I will now turn to a discussion of George Schaller's book and his work in the region.

The Stones of Silence

George Schaller came to Pakistan in the early 1970s with the primary aim of studying wildlife in the Chitral Gol area, a princely hunting reserve maintained by the rulers of Chitral. Schaller was interested in the endemic Kashmir markhor, a highly prized animal among big-game hunters since the era of the Raj. Schaller was thrilled to get a chance to study wildlife in Pakistan, which took him into "terrain where few foreigners had been since the British withdrew from the sub-continent over thirty years before" (1988, 5). Schaller finished his initial surveys in December 1970 and went back to the United States.

In 1972, Schaller came back to Pakistan at the invitation of one Dr. S. M. H. Rizvi. Rizvi, an important character in the elite-driven conservation history of Pakistan, was a Karachi-based eye surgeon and an avid hunter who had contributed to the decimation of the Sind ibex population in the Kirthar Hills. In 1969, he put down his rifle and became a conservationist advocating for the protection of the species. Rizvi built a drinking hole for the ibex and a wildlife viewing point in the Kirthar Hills out of his own pocket (Schaller 1988, 119). Later, in 1974, Rizvi was instrumental in having the Kirthar Hills declared the first national park of Pakistan. While conducting wildlife surveys in Kirthar, Schaller decided to survey the wild goats and sheep in the northern mountains of the country, mainly in the upper Hunza region near Shimshal.

In the early 1970s, construction of the Karakorum Highway had just progressed up to Hunza, so access to the area had become relatively easy, but there was still a ban on foreigners traveling to the area. Here Schaller's association with Dr. Rizvi came in handy; Rizvi obtained permission for him directly from Prime Minister Bhutto (Schaller 1988, 79). During his surveys of wild ungulates in the upper Hunza region, Schaller described the land-use practices of the Wakhi pastoralists, most notably the Shimshalis, with disdain, characterizing them as a threat to wildlife.

Schaller had previously studied other ungulates in the region, in particular the Nilgiri Tahr in India and Bharal in Nepal. His studies of these ungulates' distribution and adaptation to extreme environments, from the subtropical hills of Kirthar near Karachi in Sindh to the moist temperate hills of the Western Ghats to the heights of the Himalayas, Hindukush, and the Karakorum and the steppes of the Pamir and Tienshan Mountains, led him to construct a transnational ecology of ungulates (Lewis 2003). If the geographical scale of Schaller's studies was continental, the temporal scale of his theoretical speculations was geological. In this vast space of global conservation, Schaller constructed wildlife areas as remote pockets of refugia. Schaller opened his book, *Stones of Silence*, which was based on his fieldwork on mountain ungulates and became the basis for the establishment of the KNP, with the following:

At these heights, in this remote universe of stone and sky, the fauna and flora of the *Pleistocene* have endured while many species of lower realms have vanished in the uproar of the elements. Just as we become aware of this hidden splendor of the past, we are in danger of denying it to the future. As we reach for the stars we neglect the flowers at our feet. But the great age of mammals in the Himalaya need not be over unless we permit it to be. For epochs to come the peaks will still pierce the lonely vistas, but when the last snow leopard has stalked among the crags, and the last markhor has stood on a promontory, his ruff waving in the breeze, a spark of life will have gone, turning the mountains into stones of silence. (1988, 1)

Schaller defines the region as worthy of a national park by referencing two main conditions: untouched landscapes ("lonely vistas") and the presence of rare wildlife (the elusive snow leopard). This quote is similar to and yet different from Aldo Leopold's famous words in *A Sand County Almanac* (1949) describing his grief at shooting a gray wolf: "We reached the old wolf in time to watch a fierce green fire dying in her eyes. I realized then, and have known ever since, that there was something new to me in those eyes—something known only to her and to the mountain" (1987, 130). This quote perhaps brought about a paradigmatic shift in thinking on nature conservation in the United States and beyond.

The difference between Leopold's and Schaller's quotes is that of period and hence of context of the conservation movement. Leopold is celebrating enlightenment while Schaller is lamenting ignorance. Leopold learns about the existence of an independent nature and a fundamental principle of ecology in the process of driving wolves to extinction in the cause of agricultural expansion. Schaller is mourning the fact that we have driven animals such as the snow leopard and markhor to extinction and have not learned anything. The similarity lies in the structural elements of the two passages. Leopold shows that there is an ecological relationship between the mountain ecosystem and the wolves, but he does not tell us, at least here, what will be the consequences of severing that relationship. About four decades later, Schaller is reminding us that the mountains will turn to "stones of silence" when this relationship is destroyed. But Schaller also reminds us that it is the remoteness of the region, away from human pressure, that makes possible the survival of the wildlife still here. So setting aside areas that are already poorly connected to mainstream society perpetuates their remoteness and economically marginal position.

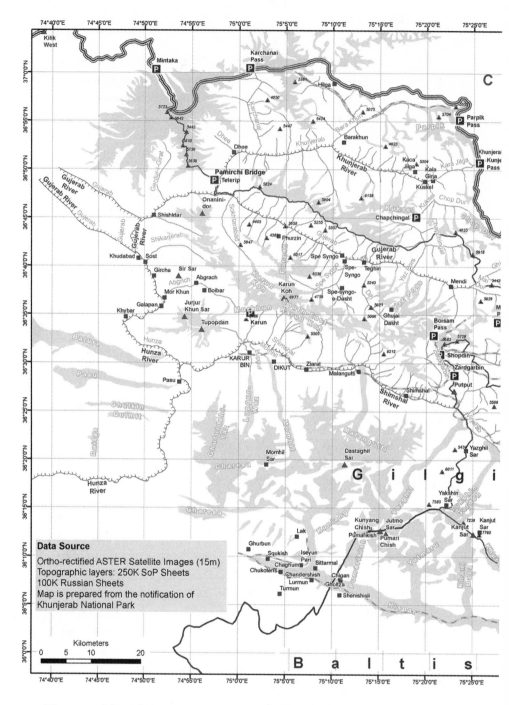

Figure 13: Map of the Khunjerab National Park, Pakistan (© GIS Laboratory, WWF–Pakistan, 2013)

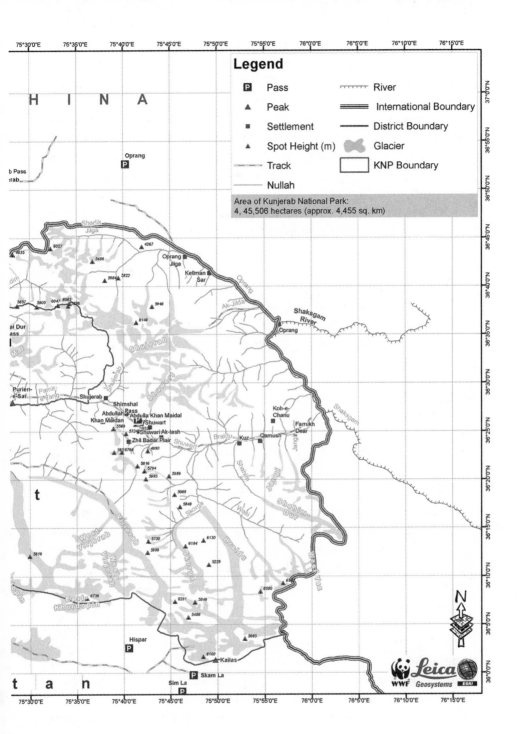

According to Schaller, human activity in the region was destroying in a few decades what had taken nature epochs to produce/evolve. It is interesting to compare this contemporary warning of *a threat from society to nature* with the late nineteenth-century discourse in which the remoteness and isolation of a society from other human societies was seen as *a threat to society from nature*; nature being the condition of the wild, unsocialized Hunza savages. But now, what was the desolate and unforgiving landscape of the days of Younghusband and Curzon has become pristine nature, and the cultivated villages of mountain subsistence farmers have become a destructive force. For example, Schaller writes: "I had ventured into the mountain naively, thinking that I would be penetrating one of the last greatest wildernesses. But seeing that here, too, man had become a destructive parasite upon the land, I became as much concerned with conservation as studying wildlife" (1988, 100). And again, "To me the most startling discovery was the extent to which the mountains have been devastated by man. Forests have become timber, slopes have turned into fields, grass has vanished into livestock and wildlife into the bellies of the hunters. The future of some animals and plants is now in jeopardy" (97).

Notice the imagery of plunder and devastation by the people of Hunza Schaller evokes. This imagery is very similar to representations of Hunza society made by British colonial officers at the height of the Great Game. Schaller was aware of this historical representation of Hunza and wrote admiringly about the role the British played in bringing order to the local society. Referring to the caravan raiding for which the people of Hunza had become known, he wrote, "Being orderly people, the British could not tolerate such restless behavior among people they ostensibly governed" (1988, 34). This orderliness that Schaller admired was one in which the people of Hunza were made less threatening—not to nature, as Schaller wants to see, but to society. So, although Schaller sees continuity in the disorderly behavior of the people of Hunza, there is an important disjuncture over time with regard to the relationship between Hunza society and nature. The British saw in Hunza society savage people living within a savage nature and as such a threat to the society of British India. Schaller sees the Shimshalis of Hunza as a society existing outside an ennobling nature and therefore as a threat to it. In the British discourse remoteness and nature were negative categories, but in the modern conservation discourse of Schaller remoteness and nature had become positive categories.

Schaller argued that local grazing practices had severely depleted the Marco Polo sheep population, and based on that, he recommended that almost all of Shimshal's grazing area be declared a national park (1988, 94). "The fact that several communities graze their livestock for about three months each summer at Khunjerab, Shimshal, and other upland areas poses some problems, for by definition a national park should be free of such *disturbances*. However, I felt that *such details could be resolved later*" (99; emphasis added).

On April 29, 1975, on the recommendation of Schaller, Pakistan's prime minister, Z. A. Bhutto, declared the Khunjerab area as Pakistan's second national park. Here again the resourceful Dr. Rizvi was instrumental in pushing the conservation agenda in the official corridors of power.[7] According to my Shimshali informants, Bhutto knew nothing of Shimshal's history and geography because of its remoteness. This remains particularly painful for the Shimshalis because they were all Pakistan People's Party supporters. Schaller was aware of the political nature of conservation and the economic burden that it puts on local people. But he was unaware that the Shimshalis used the area all year round and not just for three months of a year; that is, their winter pastures were also included in the proposed park area. This fact is not mentioned in his book.[8] Four decades on, the problems caused to the Shimshalis by the creation of the national park still persist.

Communities in Nature

I will now look at another type of nature conservation discourse in which people are seen as belonging to and being part of nature. Within the context of environmental conservation, this discourse emerged in the 1980s as a response to the failures of nature conservation discourses that separated humans from nature. This discourse on nature conservation became popular among conservation NGOs, urban activists, and state institutions. According to this discourse people in remote and rural areas are not anticonservation; rather, their cultural practices and environmental knowledge have survived uncontaminated by modernity but now need special protection against that modernity (Conklin and Graham 1995; Hirtz 2003).[9]

Using the case of the Shimshal Nature Trust (SNT),[10] a nature conservation organization established by Shimshalis in the 1990s, I examine how

Shimshalis draw on this increasingly popular discourse[11] to develop a particular strategy of resistance to the KNP regulations. Shimshalis engage the Pakistani state, sympathetic researchers, and the general public by representing themselves as caricatures of "ecological noble savages" (Redford 1991), who have survived in the remote recesses of the Karakorum Mountains and whose cultural practices are in sync with nature.[12] In the process, Shimshalis appropriate and rework the very same environmental categories of the protected areas–based approach they are resisting. How far these performances and practices are effective in changing the balance of power in favor of the Shimshalis is difficult to assess. Such performances and strategic essentialism are actually tactical essentialism because they are always deployed in response to the discourse, and hence the strategy, of other, more powerful players. Under such conditions a less powerful player, such as the Shimshalis, can make only tactical, short-term, ad hoc, and contingent interventions. There is, however, hope in all forms of subversion and resistance tactics; their accumulation over time could lead to strategic gains. Let us now turn to the story of how Shimshalis engaged with an external discourse in which their geographical remoteness and their relative lack of contact with modernity are positively evaluated.

As I discussed in the last chapter, throughout the 1980s a discussion took place on the pages of the guest book in the Shimshal hotel among Western tourists regarding how the Shimshalis' remoteness was ending and how they should embrace modernity. During this time, another important discussion was taking place in Shimshal. While the tourists had asked how the Shimshalis should embrace modernity, environmentalists were asking what modern society could learn from the Shimshalis' wisdom and environmental practices that had survived untouched by modernization.

In the early 1990s, many environmental and development NGOs in Pakistan, such as AKRSP, WWF, and IUCN, held these discussions, creating images of sustainable communities, especially in the mountainous north, as the last bastion of dwindling environmental knowledge. These discussions resulted in a new conservation policy broadly known as the community-based approach to conservation, markedly different from the protected areas approach espoused by Schaller and others. The NGOs often promoted their interventions on the grounds that they were working in partnership with already existing institutions of local resource management that were sustainable but under threat,[13] in need of support and protection. In addition to being popular in the domain of environmental conservation, such images of ecological noble savages became popular in the international media, which attempted, generally speaking, to empower them.[14]

Munawar, my Shimshali friend, recalled one summer afternoon in 1987 when he met Hideki Yamauchi, a Japanese photojournalist, on the Passu-Shimshal trek. Munawar describes his encounter with Hideki as a life-changing experience for him and many of his fellow Shimshalis. Hideki had come to Shimshal disenchanted with what he saw as the meaninglessness of life in Japan's modern society. In an introduction to one of the research reports that he coproduced with the people of Shimshal (see below), he states, "Our modern civilization has relied on physical development and physical prosperity. But now, this monstrous physical civilization starts to ruin itself, starts to spoil us, not only human beings but also all creatures of this planet" (Symposium of Shimshal Environmental Education Program 1999, 1). This kind of lamentation has a long history in different periods of Western societies (Williams 1973; Worster 2006). As we have seen throughout this book, the region has been long implicated in a durable Western discourse of decay at the center of civilization and rejuvenation in remote regions. But now this tension was being played out in a different environment and cultural domain.

In Japan Hideki worked for a new local TV channel making documentary films. His task was to find a people whose lifestyle represented a "real" experience with nature and whose knowledge was based on practical wisdom. Hideki wanted to introduce to the Japanese people a culture fully in sync with its environment where people lived a salubrious and healthy life. "He was looking for such a culture to introduce a 'paradigm shift' in Western civilization," Munawar told me. When Hideki came to Shimshal, Munawar and his cousin Inam (now working for an international donor in Islamabad), were young college students. Hideki asked them to teach him all about Shimshali culture, especially their rituals and festivals, folk philosophy, and their environmental knowledge. Hideki eventually started an exchange program between the village of Shimshal and Nihon University in Japan. During the early 1990s under the exchange program, Japanese undergraduates from the Department of Environmental Studies and Anthropology visited Shimshal and participated in daily life, learning the Shimshalis' practical knowledge. Shimshali students gathered data on the environment, local systems of natural resource management, rituals, festivals, and calendars.

The results of this research were recorded in books, videotapes, and so on that were made available to outsiders as well as to the Shimshalis themselves, to preserve their way of life from being lost. Under the exchange program, Munawar and Inam visited Japan on three occasions. They learned Japanese and lectured on the topics of practical environmental knowledge and the virtues of folk wisdom and family life on Japanese television and in local schools. In Shimshal, under the exchange program, local Shimshalis were given the title of "professor" of various subjects. The most experienced shpun[15] was given the title professor of livestock and animal husbandry, Master Daud Ali was given the designation of professor of history, and Money Beg, the most formidable hunter in the village, was given the title of professor of wildlife.

Munawar remembers that a new world opened for him as he worked with Hideki. "Hideki made us realize that what Shimshalis have is worth fighting for and modernization and development is nothing but a mirage," Munawar told me one afternoon over coffee at the posh Serena Hotel in Gilgit. I had come to Gilgit to look for hunting records of the colonial officers of the Gilgit Agency and bumped into Munawar, who was in Gilgit at the time. He said that Hideki made him and other Shimshalis realize that true and real happiness is found in one's own culture and surroundings. Shimshal emerges in Hideki's writings as a place where people cherish family ties, personal relations, social bonds, and affinity to a specific place. According to Munawar, a new and different type of consciousness about the self and culture started to emerge among the Shimshalis. They began to see their own practices in a new light, comparing point for point their own system with the West's. "To Shimshalis our culture and tradition began to appear as equalizing forces, forces that put places like Shimshal on an equal footing with the West. The West may be technologically rich, but it is culturally and ecologically pauperized," Munawar told me, his sincerity obvious.

I had known Munawar long enough to suspect there was a strategic element to this local discourse. Munawar is a highly educated, intellectually driven, and thoughtful person with immense experience of community and institutional development. He acknowledged the strategic character of this discourse but only as a matter of fact unrelated to its theoretical significance. He said that the most clear example of this strategic thinking was the establishment of the Shimshal Nature Trust as a challenge to the state's directorate of KNP.

The Shimshal Nature Trust

In the mid-1990s, a few educated young men from Shimshal who had been part of the Nihon University exchange program decided to set up the Shimshal Nature Trust. The most important body of the SNT is its task force, comprising most of the educated and professional men of Shimshal, including Munawar and Inam. All of them were living outside Shimshal—in Gilgit, Islamabad, Karachi, Lahore—but were regular visitors to the village. These were the "tribal elders"[16] (Tsing 1999) of Shimshal. In my discussions with the SNT members, they talked about environmentalism in nostalgic terms, referring to childhood memories and their experience of the "natural life," but also in terms of more progressive and modern categories such as "practical environmental knowledge," drawing on rhetoric prevalent in national and international urban centers. It was through the platform of SNT that Munawar and Inam thought of articulating their cultural survival and preservation of local knowledge as a strategy to oppose the imposition of the KNP. Munawar, one of the founding members of SNT, recalls that after their exchange program with the Japanese, the young men realized that their knowledge and culture were tied to the movement of the yaks, which in turn was dependent on the availability of the pasture on which the government had established the KNP. Munawar recalls, "We told the elders of the village that we cannot fight the government physically; the only option we have is to fight with the pen."[17] Munawar later elaborated on this, by which he essentially meant articulation of local cultural and global conservation discourse.

Under the exchange program, Japanese students helped the Shimshalis form "research committees" comprising Shimshali schoolchildren. The purpose of each committee was to collect data on a particular subject, record them, and then make a presentation to the village. The following research committees were formed: Flora and Fauna Committee, Earth Science Committee, Population Committee, Daily Life Committee, Livestock Committee, Tourism Committee, Festival and Ceremony Committee, Politics, Economy, and Society Committee, Weather and Climate Committee, and Agriculture Committee. The research produced by the committees became the basis for a pasture management plan (MP) for the whole of Shimshal later on in the 1990s.

The MP emphasized the issues of sustainability, carrying capacity, and local means of monitoring pasture conditions. A special section was devoted to the sustainable use and conservation of wildlife. In the MP, and in the research produced for Hideki, the names of animals were recorded in Urdu, Wakhi, and Latin. Likewise, flora diversity and its distribution were documented according to ecological zones and cultural zones. Both flora and fauna were then described in terms of their significance for a particular festival and ritual. In this research, then, ecology and culture emerge as a single unified and interknitted whole. The MP based on this research was eventually turned into a fifteen-year plan touted as an alternative to the government-sponsored management plan of the KNP that was produced in 1995 (Shimshal Nature Trust 1994–2009).

For indigenous knowledge to be become valuable or even decipherable to the outside world, it must be represented in modern bureaucratic categories (Hirtz 2003). Indigenous people[18] must become aware that their knowledge is of interest to outsiders and must learn how to present that knowledge in an appealing way (Brosius 1997). Although not identified as "indigenous people," Shimshalis were able to use the same strategies because of their prior experience of working with development organizations like AKRSP, which had "taught" them the art of social organization, creating linkages with other institutions, and communal management of natural resources. Munawar had worked for AKRSP and had experience overall of Ismaili institutional support to Shimshalis. But Munawar told me that it was not only the Shimshalis' strategic performance that was important in sustaining this discourse but also the desires of Hideki. He said, "Shimshalis became what Hideki wanted them to become, but Hideki also saw what he wanted to see in us." Manawar acknowledged the performative nature of his interaction with Hideki, but he did not see it as manipulative. Perhaps, as others who have theorized on such behavior have argued, Munawar also was conscious of the fluid nature of identity such that it can be deployed strategically depending on the circumstances. In the context of environmental struggles, there seems to be a general consensus among anthropologists on this point (Tsing 1993, 2004; Li 2000; Brosius 1997; Ferguson 1999; Conklin and Graham 1995).

Ostensibly produced by the SNT members themselves but in fact written primarily by David Butz, a Canadian geographer who now teaches at the University of Brock in Ontario, the SNT MP presents an image of Shimshalis to the outside world that is particularly directed to the conservation community and the national government, although among its intended audiences are also foreign trekkers, multilateral donors, and the environmental media. Written in a Shimshali first-person voice, the document states: "Although we are eager to enjoy the benefit of increasing access to the outside world, we feel a strong obligation to preserve our unique cultural environment. As a member of Pakistan's few mountain communities that retains a strong commitment to a surplus oriented economy based on transhumance livestock herding and agricultural production, we also retain beliefs, knowledge, and practices relating to nature that have been lost elsewhere." Further, "Despite a strong and responsive local ethic of conservation and stewardship, we fear that changes wrought by the completion of the road . . . and the increasing orientation of our youth toward Pakistan's urban core, will result in degradation of our natural surroundings, and the loss of our culture" (Shimshal Nature Trust 1994–2009). Notice the similarity between the above quotes with the main gist of the debate in the guest book about the coming of the road to Shimshal discussed in the last chapter. The self-image displayed is that of an "eager" people who are open to the world but also committed to their own traditions: Shimshalis represent themselves as a society that is a perfect and unique hybrid of modernity and tradition. This hybridity, however, is not seen as a mixture—rather, it acquires a uniqueness and singularity (Kopytoff 1986). Thus, the Shimshalis mobilize a unique cultural identity within the antimodernist romantic environmental discourse of Hideki. They represent themselves as the last standing culture against the standardizing wave of modernity when they state that cultures similar to theirs have been lost elsewhere.[19]

Notice also the emphasis of the Shimshalis on surplus production. It should come as no surprise that the Shimshalis attempt to represent themselves as a surplus-producing community rather than a poor, marginalized, subsistence-oriented people who need natural resources simply for their survival. One of the most important claims that the Shimshalis have been able to make is that their control of resources is important for the survival of their culture and identity, *not* for their economic well-being. The emphasis on economic versus cultural survival is important within the context of a global discourse that privileges indigenous cultural survival. Munawar told me one day, "Economic status and well-being is subject to 'natural cycles' of the markets. One can lose and regain economic status and well-being, but cultures cannot be regained once lost." The threat to culture, as opposed to economic well-being, because of its irretrievability, sends a far more potent and powerful message to the popular audience.[20]

Another important aspect of the SNT MP is its emphasis on local cultural practices, which it claims are complementary to and in line with modern conservation principles. "Historical events are remembered in detail in songs and stories, and some parts are re-enacted in skits at community festivals. These provide us with guidance for the appropriate stewardship of our landscape, and infuse our practices with meaning and an ethic of conservation, which is strengthened by a more general Islamic religious ethic of nature stewardship and respect for nature as God's ultimate creation." The document continues: "In the village, game was hunted and eaten only on ceremonial occasions, or to feed volunteers working on community projects. Hunters were admonished to shoot only the oldest animal, only one per hunting excursion, and never pregnant or nursing females. Indeed one of the most treasured songs recounts a sad conversation between a baby ibex and her dying mother" (Shimshal Nature Trust 1994–2009).

These are all examples of articulation of a certain environmental identity and positioning (Li 2000; Tsing 1993) within the modern environmental discourse. For example, the idea of a responsible hunter who takes only an old male is in line with the conservation principles of a scientific offtake or sustainable harvesting regime. Shimshalis also explain how they have divided their pastures into different land-use zones, claiming that this zoning system is in line with IUCN category 4 national park regulations. For example, they assert that the wilderness zone of the national park matches with their conception of the mergich realm, as discussed in chapter 6, which, according to Shimshali belief system, is occupied by spirits who control the wildlife. They maintain that any human activity in this zone is prohibited by their cultural beliefs, thus creating congruence between cultural and conservation sanctions about resource use.

The tone of the SNT plan aims to demonstrate to government officials and the wider world that the Shimshalis' traditional land-use practices sanctioned by cultural norms are as good as, if not better, than the principles of conservation enshrined in the national park regulations. "Our largest challenge is not to develop a system of utilizing the natural surroundings sustainably, rather to express our indigenous stewardship practices in a language that will . . . persuade Pakistani authorities that we are indeed capable of protecting our own natural surroundings" (Shimshal Nature Trust 1994–2009). Based on the claims they make in the SNT MP, the Shimshalis have made repeated requests to the directorate of KNP that the management of the park should be handed over to the SNT. Recently, Shimshalis have moved to register all their pastures—that is, land within the national park—under the ownership of the SNT. One of the objectives behind the establishment of SNT was to have it registered as a trust under whose name the entire land could be registered. Although they argue their case in terms of their unique culture, Shimshalis want to remain connected to the abstract category of the nation-state of Pakistan. Perhaps because of their distance from the center of power, it is in their interest to reaffirm their allegiance to the state even as they challenge its policies. Indeed, the reference to Pakistan and Islamic tradition reveals the power of the idea of the state among Shimshalis. They know that because of their position at the geographical margins of state control, their discursive articulation with the state is more useful than harmful. At least by affirming their political allegiance to the state, the Shimshalis are dispelling—or at least not invoking—suspicion on the part of the state. Also, the state is an important source of resources and development assistance that Shimshalis tap into when opportunities arise. But despite this strategic essentialism and articulation of discourse, the state's behavior remains unpredictable and not responsive to the aspirations of the people.

The Flight and the Fight of Shimshalis

In the summer of 2005, I attended a meeting between the office-holders of the SNT and the director of the KNP and a senior Forestry Department officer from Astor (Gilgit-Baltistan). The director was not amused by the Shimshalis' demand that the management of the KNP be handed over to the SNT. Irritated, the KNP director told the group of Shimshalis such a demand was totally out of order and that they were turning into the Bugtis of the Northern Areas. On the surface this was a rhetorical remark, but it carried an implied threat of real and material violence against the Shimshalis. The Bugtis, one of the largest tribal groups in the southwestern province of Balochistan, are engaged in a nationalist conflict with the state of Pakistan. In 2006, their leader, Akbar Khan Bugti, was gunned down by government forces and hundreds of his supporters were thrown in jail.

Comparing the Shimshalis to the Bugtis was a not-so-subtle reminder to the Shimshalis that they were challenging the writ of the state. In the current geopolitical instability of the region, the Pakistani state is jittery about its status as a sovereign power; any marginal group resisting state policies could potentially provoke unrestrained state violence. The director of the KNP was indeed making a veiled threat.

In response, Zameem Shah, president of the SNT, told the director that if the government insisted on imposing the park on their pastures, then it might as well treat the Shimshalis as part of the wildlife in the park, too. "If the national park is established, we will flee to the mountains like markhor and ibex and live with them. After all these years we have learned to live with them in harmony and it is only fair that if the government is concerned more about the fate of the wildlife than that of humans then humans will turn into wildlife."

In representing themselves as wildlife to the director of the KNP, the Shimshalis create an ambiguous self-identity, which could be either as dangerous and threatening as wildlife or as nonthreatening and endangered as wildlife, playing on the dual paths of romanticism that I discussed earlier. I argue that Shimshalis self-consciously create an ambiguous identity as wild and domesticated people and mobilize opposite meanings of such representations to make their point. By representing themselves as threatened and endangered, they correlate with the official nationalist discourse on progress and modernization and seek the state's resources and protection. But their self-representation as wildlife also articulates another discourse in which they are seen as dangerous because they defy control.

The fugitive self-identity articulated by Zameem Shah at that day in the office of the director of the KNP was strategic as it challenged the most fundamental practice of the state—that is, in James Scott's (1998) term, legibility. By threatening to become illegible to escape state regulations and policies, the Shimshalis remind us that they consider illegibility to be a form of social protection built into their physical environment. By representing themselves as wildlife, Shimshalis reaffirm their representation as still wild people, part of nature, who by this very identification also have the capacity to escape state practices. They counter the predatory nature of the state by reminding it of their ability to survive such predations; they redeploy their state of nature to counter the nature of a state.[21]

Conclusion

Global interest in conservation of "wild" nature has resignified remote areas as in need of protection. Remote areas, by virtue of their distance from densely populated areas, are seen as places where little has changed in terms of physical nature. Such imaginings often fail to take into account local land-use practices and struggles over resources; even when these are recognized, they are often deemed unimportant. Modern-day conservation is directly implicated in the production of remoteness through its discourse on wilderness. Wilderness is not geographical remoteness per se; rather, it is about removing and setting aside areas from human use. The rising interest in wilderness and wild animals is turning rural lands around the world into national parks, game reserves, and game sanctuaries, thus removing them and making them "remote" from society.

This association of remoteness with wilderness in the environmental discourse, although itself more than a century old, is a continuation of seeing Hunza/Shimshal in a positive light since the conquest of Hunza at the end of the nineteenth century. Remoteness is associated more and more often with things that have escaped the juggernaut of modernity, and hence they are valued positively. The association between remoteness and wilderness also becomes possible when wild animals, too, as well as wild places, become the object of our appreciation. No longer seen as dangerous and threatening, wild animals become safe in the confines of nature reserves.

Epilogue

This book began with two vignettes of visitors to Shimshal, arriving almost a century apart, elucidating how the villagers were seen by one group as savages and by the other as an ideal, hospitable people. The rest of the book showed how the region and the people of Hunza and Shimshal have been imagined and reimagined during the intervening period by various outsiders, with remoteness being the constant defining theme.

The history of Hunza shows that some of the themes associated with remoteness repeat themselves, although the wider context and the specificities may change. For example, the search for an original Aryan race by the British is similar to the search for an original Pakistani ethnicity by the Pakistan People's Party. Both ascribe to remoteness the quality to harbor origins. The 1940s discourse of an idyllic rural society is similar to the discourse of hospitality in Shimshal; both construct remoteness as a refuge from modernization but in the process affirm that very modern structure of feeling that makes the escape enjoyable. Understanding these different themes, often contradictory or inconsistent and sometimes mutually reinforcing, requires us to place the concept of remoteness in a wider history of ideas and paradigms and political processes associated with them.

If distance, both geographical and sociological, is the defining feature of remote areas and remoteness, as has been shown in this book, then how does this condition affect the ways in which they are understood and represented by outsiders? Perhaps a more pertinent question theoretically would be to ask what "methodology" is used in the modern world to understand distant events and processes, phenomenologically as well as cognitively. This history shows that remote areas are a part of the imaginative process in the modern era, and of the human conditions associated with it.[1] This imagination is not to be taken here as a work of fiction or a flight of imagination but as a way of understanding the world.

The earliest to pay systematic attention to this question was Charles Wright Mills, who emphasized the role of the imagination in conceptualizing the modern and postmodern world. According to him, the modern world had become complex to the point of the moral breakdown of society.[2] He argued that understanding this complex modern world—he was writing about the human condition in the second half of the twentieth century—is essential for any meaningful engagement with it. We need "sociological imagination" to understand how the modern world is affecting us so that we can take appropriate actions.[3]

Commenting on a similar problem about the postmodern age, the Marxist literary critic Fredric Jameson develops the concept of cognitive mapping as a means of using the imagination to grasp the "true" totality of the world in which we live, a grasp essential for any meaningful political action. Jameson's goal is for people to be able to carry in their heads a "mental map of the social and global totality," much as city dwellers are able to do for the cities in which they live, so enabling them to bring together both the "here and now of immediate perception and the imaginative or imaginary sense of the city as an absent totality" (1988, 350).

Arjun Appadurai, grappling with a similar problem in *Modernity at Large*, considers the role of imagination in conceiving of a totality larger than individual experience. He argues that "electronic mediation and mass migration mark the world of the present not as technically new forces but as ones that seem to impel (and compel) the work of the imagination" (1996, 4). In the postmodern, postelectronic era, Appadurai argues, imagination has to play a significant role in the lives of ordinary people the world over. The spread of global mass media and immigration has made imagination a social practice that leads to the formation of new concepts of the world and collective identities beyond the local and regional domains, thus making modernity "at large" (3–4).

The theoretical propositions made by these three scholars show that imagination is at the heart of the social construction of the modern world. Politically, socially, and economically remote areas are encountered at the edges of modernity and its processes, that is, they are formed in the control and symbolic representation of the processes of geographical exploration, state expansion, and border administration. They are linked in an ambiguous love-hate relationship with the outside world. For example, remote areas are targets of colonial and postcolonial development projects of a conventional kind. Yet they are also appropriated by outsiders and distant players as alternative models of development. They are represented as realms of pure nature but also of authentic society. They are off the grid in everyday national politics but often have deep historical significance to the national narrative. They are made accessible only through modernization, yet they are refugia from modernization. They represent "local" places where the "global" is experienced.[4] They are also very much part of the modern experience, evoking sentiments central to that experience such as estrangement, enchantment, and revulsion. Remote areas are like empty signifiers with considerable freedom of meaning attached to them, as shown in the historically constituted representation of the Hunza people as caravan raiders (Harms, Hussain, and Schneiderman 2014, 378). Caravan raiding conjures up the danger of the breakdown of law and order, but it also evokes natural freedom. It means different things in different times and to different people.

The engagement of the people of remote areas with the outside world suggests that they are part of this wider world, not secluded from it. But they are connected to it in ways that enhance their estrangement and hence evoke enchantment or revulsion. Remote areas are not to be seen as standing in contrast to the idea of a pure cultural center; rather, they are regions that escape the orbit of dominant cultural meaning. Moreover, unlike the concept of a liminal zone, which seems to escape the dominant structure, remote areas are within the structures of modernity, but again, connected to it in a special way. The ambivalence associated with remote places suggests that the making of remoteness, like the idea of modernity itself, is never complete, as there is no settled finality of meaning that we can attach to it. Remoteness remains a real and material condition of spatiality as well as a work of imagination.

What can we make of the shifts in the identity of remote areas as we saw in the case of Hunza? What changes in global conditions of modernity bring about these changes in the meaning of remoteness? Commenting on the general process of capitalism, Fredric Jameson has charted a somewhat similar trajectory of how space has been transformed. In reference to modern Cartesian (secularized) space, Jameson writes: "We witness that familiar process long generally associated with the Enlightenment, namely, the desa-cralization of the world, the decoding and secularization of the older forms of the sacred or the transcendent, the slow colonization of use value by exchange value, the 'realistic' demystification of the older kinds of transcen-dent narratives in novels like *Don Quixote*, the standardization of both sub-ject and object, the denaturalization of desire and its ultimate displacement by commodification or, in other words, 'success,' and so on" (1988, 350–51).

We notice a similar trend in the history of Hunza. The case presented in this book shows that in the nineteenth century conceptions of territory, people, and control were different from what they are today. In the nine-teenth century, one explored remote areas to look into the possibility of colo-nization and expansion or to find lost people, tribes, rivers, and mountains because unknown territories and people still existed. The biblical narrative was taken seriously in nineteenth-century colonialism. In the first half of the twentieth century, it was the modern secular sciences of boundary making, classification of space, and industrialization that defined Hunza's remoteness. In the second half of the twentieth century, Hunza's remoteness was de-scribed in terms of nationalism, tourism, and environmentalism, all of which exploited the "unspoiled," "pristine," and hence remote status of Hunza.

This change under modernity from a religious view of space and terri-tory to a secular view, as evident in the history of Hunza as a remote area, shows a concomitant transformation in the meaning and identity of remote areas. Although, conceptually, there is no set teleology of remote areas, the case of Hunza presented here shows that from the point of regimes of gov-ernance, remote areas have undergone three types of transformation in terms of their basic identity: areas to be explored (geographically), areas to be controlled (politically), and areas to be protected (as conservation) or developed.

The imagination of remote areas is a perspective from outside, but re-mote areas are not only products of outsiders' imaginations. There is every-day life in remote areas. True, remote areas are filled with an abundance of physically empty, inaccessible landscapes and precarious social and material connections between the people who live there and between those people and the outside world, but such a context has its own habitus. As we saw, the mental structure of isolation of the people of Shimshal, based on the nature-culture dichotomy (village and pasture worlds), is being reshaped and aligned with global processes, with tourists seeking ever-more distant places to visit.

The history of Hunza also suggests that remoteness, as much as it is a fact of physical geography, is not an inherent condition; rather, it is often created. Most attempts by the rulers of Hunza to keep themselves remote, for example, had the objective of keeping themselves inaccessible from encroaching external powers. The rulers of Hunza also kept the region remote by preventing their people from leaving. During the 1950s, under Pakistani administration, the mirs actively tried to keep the areas remote because of threats to their continued exploitative rule.

Today, the people of Shimshal increasingly see their remote status as part of their strategic identity. They play with this identity to become part of the global modern world while remaining "remote." Shimshalis' yaks today carry North Face rucksacks on one side and wool, butter, and game meat on the other. Shimshalis now take cell phones to the pastures, as in some remote corners they are able to pick up signals and communicate with the outside world. The yaks' travel routes are now regularly traversed by the boots of Western and Pakistani trekkers. Old rituals of agricultural seasons are aligned with the new cycles of the tourist season. New roads open, decreasing the friction of distance but not annihilating it, giving rise to new travel conditions and new experiences of remoteness. The local transhumance economy is becoming intertwined with the global language and practice of nature conservation. The transhumance migration routes are becoming treks, treks are becoming stages, and those stages are becoming routinized, legalized, and commoditized. New words with new ideas are coming in, but the old persist.

Today, Karimabad is legendary not because of its inaccessibility but because of its sophisticated tourism infrastructure. Its culture, architecture, and scenery are marketed. It is no longer considered a remote tourist destination; culturally interesting, yes, but not remote. In Shimshal, on the other hand, it is hospitality, a precontact trait, that survives and gives the village the status of a remote destination.

In the twenty-first century, we note that certain ideas associated with remoteness have been replaced by others. Today, one can visit remote areas remotely. Google maps, GPS, and drones have not only changed the way we think about remoteness, they have also affected the way the people in these areas think about themselves in relation to others. Today, the inhabitants of North Waziristan, the site of an intensive drone campaign by the CIA, think of U.S. power as emanating from remote corners of the earth, while they— the central target of this power—are fully exposed and accessible. Trekkers and tourists sitting in Internet cafés in Prague, Berlin, London, or New

York can trace the exact route of a popular Karakorum trek near Shimshal, gathering information about the topographical gradient, water points, and the distance between stages. Thus, with changes in technology and the resultant interaction with cultural practices, remoteness today is less about the mystery of unknown distant lands, strangeness, and alterity or inaccessibility; rather, it is about the reach of anonymous power.

Today, the broader region surrounding Hunza is once again tied to geopolitics at the global level. The inhospitable and rugged mountains of Pakistan's north have once again become populated with savages and barbarians. Although Hunza is not in that region of Pakistan that is considered the epicenter of global terrorism, it is being talked about in geopolitical terms. The rising power of China in the world means that Hunza's past tributary relationship with its neighbor to the north has once again become a topic of speculation and discussion internationally. With China upgrading the KKH and possibly constructing a railway line that will eventually give it access to the warm waters of the Arabian Sea through Pakistan, there are hints of a new Great Game. In August 2010, an article appeared in the *New York Times* that, in essence, stated that the Chinese People's Liberation Army (PLA) had descended into Hunza in the thousands and taken over the region. "Many of the P.L.A. soldiers entering Gilgit-Baltistan are expected to work on the railroad. Some are extending the Karakorum Highway, built to link China's Sinkiang province with Pakistan. Others are working on dams, expressways and other projects. Mystery surrounds the construction of 22 tunnels in secret locations where Pakistanis are barred. Tunnels would be necessary for a projected gas pipeline from Iran to China that would cross the Himalayas through Gilgit. But they could also be used for missile storage sites" (Harrison 2010).

Articles such as these raise the alarm about China and its expansion into Hunza and Pakistan and indirectly pitch Pakistan as a foe of the United States rather than an ally. The article was reproduced in local newspapers in Hunza and Gilgit-Baltistan, to the consternation of readers who, through letters to the editor of the *New York Times* and the local papers that had reproduced the article, vehemently tried to dispel the impression given in it that Pakistan had handed over the region to China. The people of Hunza and Gilgit-Baltistan view such alarmist statements as akin to Younghusband's representation of them as caravan raiders, a representation that justified the invasion of Hunza and the loss of its independence forever in 1891. People in Hunza are intently watching global politics, and their own representations in it, because they realize that representations are tied to geopolitics and are followed by real consequences.[5]

Throughout history, we notice that remote areas drew their power from the fact that they escaped the reach of more powerful surrounding entities. Indeed, this is the reason that outside powers saw the limits of their authority in remote areas as dangerous in itself, indicative of the presence of a rival power that they could not control. But this fact is changing. The power of remote areas is no longer associated with their ability to evade dominant powers. Tucked away in their lofty mountains and forbidden peaks, the savages of today are no longer materially secure from distant powers. As mentioned earlier, North Waziristan, a region to the south and west of Hunza and, like Hunza, a relic of the Great Game, is now at the center of the "war on terror." It was once geographically remote and inaccessible like Hunza. But drones have changed that. Drones cannot reach everywhere, but they have reduced what can qualify as remote. Virtual technology, such as that used with drones, enables one to be "present" in a remote area even though one remains physically distant. As such, this type of virtual experience is similar to imagining a remote area without actually visiting it. However, the difference is that now those doing the "imagining" can exercise their agency in the remote area from which they are physically distant: firing a rocket from a position of imagination kills real people.

In 2010, when a major flood swept through the northern areas of Pakistan, including Hunza, a discourse of blame circulated in Pakistan, pointing fingers at the United States with claims that it had used geo-weathering technologies to remotely change the patterns of weather cycles and the direction and intensity of monsoon winds. Many people in northern Pakistan believed the flood was a man-made disaster, not a natural one. The source of the flood, a natural threat to the local society, was seen as emanating not from a threatening and dangerous nature but from a remote and distant political power, for reasons rooted in global geopolitics. Drone technology and its magical power are much talked about in Pakistan, revealing an acute sense of insecurity embedded in conspiracy theories that give free rein to the realm of imagination. Today, it is people in the classical remote regions of the past who hold fantastic representations of nonremote areas, assigning to them mythic qualities of extraordinary and anonymous violence, thanks to drone technology, satellite-guided missiles, and special ops, all deployed within the discourse of the war on terror.[6] This discourse has allowed space to be collapsed, and now no place on earth is "remote."

Notes

Introduction

Parts of this introduction's section "Remoteness and Affect" have appeared in a recent publication; see Harms, Hussain, and Schneiderman 2014.

1. See Hopkirk 1994 and Keay 1996 for popular histories of the Great Game.

2. From now on, Gilgit-Baltistan and its surrounding region to the north will be referred to simply as the region. This region is the meeting place of the Himalayan, Karakorum, and Hindukush ranges. Further north at Hunza's border, the Pamir range meets the Karakorum. This region has the highest concentration of the loftiest mountains on earth. Within a radius of one hundred miles, there are more than sixty peaks over twenty-three thousand feet. It is a spatially cutoff and isolated region, with few people and an abundance of crags and ice.

3. Hunza is also the name of the main valley of the region; however, when I use the term "Hunza" I will be denoting the wider Hunza state.

4. See Michael Dove's discussion on these two types of other. An adversarial other is imagined in terms of economic and political interests, while the alien other is considered the cultural other whose economic and other characteristics are unfamiliar (2011, 231–32).

5. I use Anthony Giddens' definition of modernity as it leaves room for individual experience with the structures of modernity at an institutional level. Giddens states, "Modern institutions differ from all preceding forms of social order in respect of their dynamism, the degree to which they undercut traditional habits and customs, and their global impact. . . . Modernity radically alters the nature of day-to-day social life and affects the most personal aspects of our experience. Modernity can be understood on an institutional level; yet the transmutation introduced by modern institutions interlace in a direct way with individual life and therefore with the self" (1991, 1). Expounding on the role of individual imagination and modernity, Arjun Appadurai goes further in arguing

that the condition of late capitalism, marked by mass electronic media and mass migration, makes it possible to understand modernity only through imagination (1996, 3).

6. See, for example, Tsing 1993 and Li 2000 on marginal areas, Hughes 2008 on frontiers, Neumann 2002 on wilderness, and Sturgeon 2005 and Scott 2010 on borderland.

7. Edmund Leach introduces us to the idea of topological space, which represents the level of connectedness of the elements of a system. Leach discusses society in structuralist terms as a figure whose shape changes at different points of connections (1961, 7–8). Thus, depending upon the connectivity, the "shape" of a culture may appear different from different vantage points. It is in this way we can think of remoteness as determined not only by topography but also by topology, that is, the level of connectedness experienced in cultural vocabulary. So two geographical locations may be equally distant in topographic space from a third location, but the connectedness of the two in physical and conceptual space may not be the same, and usually a remote place is topologically different from a nonremote one.

8. Eric Wolf writes of "remote" and "isolated" communities: "Rather than thinking of social alignments as self-determining, we need—from the start of our inquiries—to visualize them in their multiple external connections" (1982, 387). Arjun Appadurai, making the same point, states: "Natives, people confined to and by places to which they belong, groups unsullied by contact with a larger world, have probably never existed" (1988, 36).

9. One of the most important characteristics of remote areas is the singular effect of landscape on human senses and thinking, especially on the "enhanced defining power of individuals" (Ardener, 1989, 222). This phenomenological/affective character of encountering a remote area is well captured in the following description by Robert Shaw, a British trader who traveled north of Kashmir into Chinese Turkestan through western Hunza and back again in the early 1870s. "On those endless plains you never seem to arrive anywhere. For hours you march towards the same point of the compass, seeing ever the same objects in front of you. If you discover another party of travelers coming towards you in the distance, you may travel for half a day before you meet them. The air is so clear that there is no perspective, everything appears in one plane, and that close to the eyes. When, after threading these interminable valley-plains, you descend again towards the inhabited country of Ladak, the first bits of village cultivation seen on an opposite hill-side have the most singular effect. 'Cela vous saute aux yeux.' They seem to come right out of the surrounding landscape of desert, and to meet you with almost painful distinctness . . . with an atmosphere which acts like a telescope, bringing the most minute and distant objects into notice" (1871, 8). The quote reflects a sharp comparison between perceptions of traveling to and having arrived at a remote place. It describes the feeling of irrelevance and boredom of "seeing ever the same objects in front of you," and it also reflects the loss of perspective, perhaps because of the vast scale of the landscape compared to humans. But the scene of arrival at a remote place is marked by feelings the opposite of boredom. Here the author acquires a keen interest in the most irrelevant things and starts to take notice of them, leading to an increased sense of observation, the thought that somehow the social logic of this place is designed differently.

10. I use Stuart Hall's definition of discourse, which in turn is inspired by Foucault's conception of it. Hall states, "It [discourse] is a group of statements which provide language for talking about—a way of representing the knowledge about—a particular topic at a particular historical moment. It governs a way that a topic could be meaningfully talked about and reasoned about" (1992, 291).

11. See Kürti's (2001) work on how mainstream Hungarians think of Transylvania as a remote place and also as a repository of authenticity and mythical nationalism.

12. By "ethnic group," I simply mean groups that are differentiated on the basis of language and political relationship with each other.

13. In total, Ismailis represent about one-third of the total population in Gilgit-Baltistan, with Shias and Sunnis making up the other two-thirds equally. In Pakistan generally, Ismailis are a tiny minority, less than 1 percent, with the only significant population outside Gilgit-Baltistan based in the city of Karachi in the south. Most Pakistanis are Sunni (about 75 percent) or Shia (about 20 percent).

14. Moreover, other readings of the same event could also shed new light. For example, Patrick French shows that Younghusband misreported his own follies during his interactions with the local people in his published accounts (1995, 78). But such instances of alternative accounts are rare in the history of the region.

Chapter One. Lifting the Veil

1. During the mid-nineteenth century, the British increased their geographical explorations of the region that divided South, Central, and West Asia. By this time, the eastern and the central face of the Himalayan Massif had been surveyed, and toward the west, British geographical explorers and travelers were traversing the passes of the Caucasus Mountains. The northwestern edge of the Himalayas and the awesome Karakorum Range were too difficult to have been penetrated yet. Mountains rising to twenty-eight thousand feet, cut by deep and narrow river valleys with the longest mountain glaciers in the world, made it an inaccessible place.

2. James Lawrence, writing from a popular British perspective in contemporary times, likened Russian political intrigues against the British during this period to "maskirovka," a chess maneuver that basically means instilling fears in your opponent's mind that his major possessions are under threat (1997, 364).

3. Before Hunza's strategic location was fully realized, the British accorded more importance to the neighboring valley of Yasin and planned its invasion in the late 1870s. It was only after a precarious treaty was signed between the maharaja of Kashmir and the ruler of Chitral and some geographical knowledge was obtained about the passes that Hunza's importance was recognized. For a detailed and wonderful study of the settling of the frontier region, see Woodman 1969, 84–107.

4. At this time, the high mountain passes in the Karakorum, Pamir, and Hindukush ranges on the northern frontier of Kashmir were still unknown to the British.

5. I am using the word *Orientalism* differently than it is understood conventionally in literary criticism. Thomas Trautmann (1997), in his book *Aryans and British India*, identifies two kinds of Orientalism, which he calls Orientalism 1 and 2. Orientalism 1 is

the "knowledge produced by the Orientalists, scholars who knew Asian languages," and Orientalism 2 is "European representation of the Orient, whether by Orientalists or others" (23). Orientalism 1, the main focus of Trautmann's book, is, in highly simplified terms, about the exploration of Indian ancient texts by European scholars using Asian languages (mostly Sanskrit). Urs App argues that Orientalism was not born in the eighteenth century; it had a long tradition in Europe. It basically meant study of the theology of the non-Abrahamic religions of Asia such as Hinduism, Brahmanism, Taoism, and others (2009, 9). My use of "Orientalism" means Orientalism 1; hence, the East Indian Company scholars are referred to as Orientalists because they studied ancient Eastern texts.

6. The main purpose of their studies was to find the code of governance upon which a better administration of the local affairs could be carried out by the company.

7. Thomas Trautmann (1997) argues that the project of eighteenth-century Orientalists was not linguistic in character but rather ethnological. Trautmann calls the ethnological project of early Orientalists Mosaic—"that is, an ethnology whose frame is supplied by the story of the descent of Noah in the Book of Genesis, attributed to Moses, in the Bible" (41). The work of East India Company scholars in the eighteenth century remained limited to texts; the British made no effort to locate this place in physical space. It was within the context of geopolitical imperatives that British officers and explorers of the late nineteenth century took it upon themselves to find these places in actual geographical space.

8. This region of the Pamir is one of the harshest places in the world. It is known locally as Bam-e-Duniya, or Roof of the World, and is a high-mountain tableland, about seven hundred kilometers long and one hundred kilometers wide, that stretches across what is today Afghanistan, Pakistan, China, and Tajikistan. It is divided into two areas called Big Pamir and Little Pamir. Here, the British identified the Oxus as the sacred river.

9. This aestheticization of the geopolitics of finding the boundary of the empire produced what Mary Louise Pratt has called the "anti-conquest" narrative. She describes it as "strategies of representations" of the innocence of European visions of their practices (1993, 7, 39).

10. Clearly I am simplifying here, as one cannot separate the application of a discourse from its production. Here "lifting the veil" also carries gendered connotations, as in "penetrating virgin land." The British also framed their conquest of colonial lands as the sexual conquest of exotic females. This point is perhaps illuminated by Edwards (1989), who points out that the expression is "veil" rather than the more conventional term geographical "curtain."

11. I argue that the metaphor "blank on the map" rhetorically displaced the region epistemically—that is, its knowledge was presented as nonexistent because of its remoteness. The metaphor of "blank on the map" set the stage for the extraordinary character of the place, and then the use of metaphors such as "lifting the veil" placed the region on the map.

12. During the second half of the nineteenth century, rising bourgeois work ethics had created a cultural discourse in which endurance and performance became new

ideals for the British. Indeed, sports played a big role in inculcating this ethic (Mangan 1986). For a wonderful study of the influence of ideas of courage, competition, endurance, and risk in British mountaineering in the Himalayas, see *Fallen Giants: A History of Himalayan Mountaineering from the Age of Empire to the Age of Extremes* by Maurice Isserman and Stewart Weaver (2008). Exploration of remote areas provided an opportunity for the affirmation of these ideals as much as it offered avenues for thrills, riches, and fame. Writing about British explorations in South and Central Asia, Geoff Watson states that the difficulties of obtaining knowledge about territories beyond the boundaries of the British Empire in a place that posed serious dangers and stern physical challenges meant that exploration of Central Asia was presented as an affirmation of British prowess. The region also attracted explorers, tourists, and missionaries because of its isolation and exotic reputation (2002, 149–50). For most of the nineteenth century, exploration in Central Asia was a masculine endeavor, the majority of explorers having a connection with the military and the Royal Geographical Society.

13. A. L. Basham wrote, "The modest spherical earth of the astronomers did not satisfy the theologians, however, and even later religious literature described the earth as a flat disc of enormous size. In its center was Mount Meru, around which sun, moon and stars revolved" (1959, 488).

14. Tom Neuhaus (2012) has described a similar process taking place in the case of Tibet during the nineteenth century, when Tibet was beginning to enter the Western imagination as a "veiled" place.

15. A trip to the source of the river Oxus from Calcutta via Afghanistan rather than Hunza, as became possible later on, could take up to two months if one moved with full speed, given the available modes of transportation at the time. Often such exploratory trips lasted six to eighteen months, which also sometimes included periods of detention or arrest.

16. It is not clear if Macartney actually visited the region or, like Elphinstone, relied exclusively on native informants. Curzon's (1896, 52–53) commentary on Macartney's description of the source of the river Oxus suggests that he may have gathered this information from his native informants. There is no information on the route that Macartney took from Peshawar to Kabul and then to the northeastern region of Badakhshan where the Oxus rises.

17. This region of the Pamir Mountains is a high tableland, ideal for grazing during summer. In the local Wakhi language, "Pamir" means an open upland valley. This region is divided into two parts, Little Pamir and Big Pamir.

18. The native spy is an interesting historical ethnographic subject about which not much has been written.

19. Douglas Forsyth's first mission visited Kasgharia in 1870.

20. This commission was responsible for demarcating the boundary between Afghanistan and Russia; the former belonged to the British sphere of influence and the region north of it to Russia's.

21. Although the author of the report seems to agree with Curzon's claim, the actual boundary was set on Lake Victoria, which was claimed to be the source of the Oxus by Wood.

22. Watson argues that the knowledge produced about Central Asia went through three phases, as described by Joseph Conrad:, "By the first, 'Geography Fabulous,' Conrad meant 'the fantastic visions of mediaeval cartography.' The second was 'Geography Militant,' paraphrased by Driver as 'a rigorous quest for certainty about the geography of the earth.' The third was 'Geography Triumphant,' which depicted a condition as the number of unknown spaces in the world diminished until no area was left unexplored" (2002, 101). Mathew Edney has also described a three-phased development of cartographic representation of India between the sixteenth and eighteenth centuries (1997, 4–5).

23. Recall that the Royal Asiatic Society of Bengal was established by none other than Sir William Jones, the Orientalist scholar of the East India Company in the early nineteenth century, for studies of the ancient Hindu texts and traditions.

24. About these sources, Clark writes, "The general reason these sources are quoted in current works is not the information they give on the people concerned but because they provide an introductory pedigree to a study. It is conventional to quote them; they are a scholarly reflex, used because they have been used. What is sociologically more interesting is that these classifications themselves have the potential of 'creating' the very people they apparently describe. In this way the classification can become a self fulfilling prophecy" (1977, 333).

25. By this treaty, Kashmir became a vassal state of the British Empire. This relationship was acknowledged in the shape of Kashmir's annual payment of tribute of Kashmiri shawls to the British (Chohan 1985, 2).

26. Nagar was a neighboring state located east of the Hunza River.

27. For a wonderful discussion on the political and military origins of the term *Dard*, see Graham Clark's (1977) "Who Were the Dards?"

Chapter Two. The Friction and Rhetoric of Distance and the Alterity of Hunza

1. Scott states, "We could say that 'easy' water 'joins,' whereas 'hard' hills, swamps, and mountains 'divide'" (2010, 45). According to Scott, hard to get to places are state-resistant spaces because their difficult terrain and topography enable them to avoid the reach of centralized states.

2. C. E. Biddulph, writing in the late nineteenth century, criticizes accounts of explorers, which he thinks are based on the authors' "imagination": "The ground has been prepared for such exaggeration on their part by the daring hyperbolisms with which the various Orientalist writers on Central Asian matters . . . depicted the wonders and magnificence to be found there" (1891, 563–64).

3. Just as Tartary or Siberia offered images of the exotic that Russians mobilized in creating images of the self, so Ezo was a source of visions of otherness, which fed a slowly emerging consciousness of Japan as a nation. Perceived (following the Chinese model) as a "barbarian periphery," Ezo was at first depicted in language full of magic and the monstrous. A fourteenth-century Japanese scroll described the "thousand isles of Ezo" (Ezo-ga-chishima) as inhabited by cannibals, shape-shifters, and female demons (Morris-Suzuki 1999, 65).

4. There is no truth to this claim.

5. Similar representations were made of neighboring tribes. For example, about the Kalash people, George Scott Robertson, the surgeon general of the Gilgit Agency, wrote that they were a race of "thieves" (1896).

6. Notice that when Younghusband traveled through the region in 1889, the road leading north from Srinagar to Chinese Turkestan through Gilgit and Hunza was in very bad shape. It was because of this that Younghusband set the boundaries of civilization in Srinagar. About three decades later, when the Srinagar-Gilgit road had been built and heavily used, Jenny Visser-hooft, a Dutch geographer, wrote, "Beyond Gilgit civilization ends" (1925, 29). So the construction of the road into this region symbolized the penetration of civilization into an otherwise civilizationless area because of its remoteness.

7. As in Zomia, as described by James C. Scott (2010), the friction of distance added to the ease with which local states remained on the margins of imperial states. The Russian frontier had moved steadily eastward with the building of roads and the construction of railways. The Persian Empire had also connected hinterlands with the regional and central institutions of power. Remote areas were increasingly coming under the orbit of imperial design, and Hunza was no exception.

8. Biddulph expresses admiration for the surefootedness of local people when he writes about the difficult terrain, "The roads are of the rudest kind, and necessity has made the inhabitants intrepid cragsmen; they pass with ease over places so dangerous that even experienced mountaineers would frequently hesitate to follow them" (1880, 2).

9. For detailed accounts of the events leading up to the establishment of the Gilgit Agency in 1889 and the subsequent invasion of Hunza in 1891, see Woodman 1969 and Chohan 1985.

10. Plowden to secretary of state, Government of India (GoI), September 9, 1886, R/2/1079/246.

11. Plowden to secretary of state, GoI, May 9, 1888, R/2/1079/245.

12. H. M. Durand to Colonel Nisbet Parry, May 22, 1889, R/2/1061/1–8.

13. There are two routes from Srinagar to Gilgit: Burzil and Kamri Pass. Durand preferred the Kamri Pass route because it was shorter and passed through an area with abundant forage available for the pack animals, but this route was in a horrible condition beyond the pass in 1889, so the British ended up using and improving the Burzil route (Landsdowne to secretary of state, GoI, May 6, 1889, R/2/1061/1–8).

14. From Baltit to the northern frontier of Hunza with Russia and China, at the source of the river Oxus, was another 110 miles, a terrain most unsuitable for any mode of travel other than foot.

15. The local headman of the village, or *numberdaar*, was to provide food items at controlled rates to the official travelers of the Kashmir state. A special *parwanah* (an official slip) was issued to any official travelers by the Kashmir state, upon the presentation of which local people were obliged to provide food and labor.

16. This was the name that was rejected by Leitner and replaced by Dardistan.

17. For example, in 1875, Frederic Drew stated, "Another route leads by a somewhat difficult pass on to the Shimshal Pamir, from whence a road goes to Ujadbai in the

Sirikol Valley. On this Pamir dwell a number of Kirghiz, who pay tribute to the Tham of Hunza" (1875, 26). Another example comes from a report written by Lieutenant Manners-Smith, a military officer of the Gilgit Agency handpicked by Durand. Durand had sent a Chitrali prince to spy on the Russian movement in the Oxus region north of Hunza in 1890, and he reported on what the spy told him: "The Kirghiz version of Raja Safdar Ali's [mir of Hunza] attack on the Kirghiz of Taghdumbash is, that the headman, named Kuch Mahabat, had been camping with his people at a place called Shakh Tuda, close to the Kilik Pass, and that Kuch Mahabat had enraged the Hunza raja by refusing to give the usual forced offerings of felts and dried apricots, so that the latter sent a party to coerce him" (Colonel Nisbet Parry to secretary of state, GoI, August 13, 1890, R/2/1079/248).

18. The struggle for control over the Kirghiz grazing lands was also particularly intense because of the excellent pasturage available there. With little cultivable land and very sparse pasture areas available in central Hunza valleys, the headwaters of the Raskam valleys were particularly attractive to the pastoralists and rulers of Hunza. The Earl of Dunmore, who traveled through the region in 1892 after Hunza had fallen to the British, left this impression: "They [the Kirghiz] live in their felt tents and wander about from place to place with their yaks, camels, sheep, and goats, cultivating here and there a little barley. As a rule they are tolerably well off, as they pay no rent to the Government for their grazing" (1894, 239). In a similar vein much later, Emily Lorimer nominated as the most important feature of the Yarkand River valley its luxuriant pastures. Of Shimshal, in northeastern Hunza, she wrote: "Food and pasturage are plentiful, and life in some respects more luxurious than in lower Hunza" (1939, 121).

19. Irmtraud Stellrecht, in his wonderful study of the evolution of a centralized political state in Hunza, argues that by 1800, the state had monopolized control over people and resources and the trade route between South and Central Asia (2006, 210).

20. We will discuss this point in detail in the next chapter.

21. This distinction in *styles* of killing and its reinforcement of who is and who is not civilized continues today, with beheadings of individuals by "terrorists" considered infinitely more savage than the killing of many thousands by the bombs of the "civilized."

22. I am not arguing that the mir was deploying this counterdiscourse as a calculated effort in which he had figured out all the anxieties and fears of the British and was now trying to exploit them tactically. Rather, I am arguing that his discourse was a response, not a point-for-point, to the British discourse about him. In this way, I concur with Ferguson (1999) and Li (2000), who both argue that successful performance of an identity in politics and culture is *not* manipulation per se.

23. Colonel Charles Stoddard and Captain Arthur Connolly were killed by the ruler of Bokhara, Alexander Burns was killed by a mob in Kabul, Adolph Schlagintweit was killed by the chief of Kangra, and George Hayward was killed by the chief of Yasin.

24. The Indian government regularly sent the rulers and leaders of petty tribal states to military towns in India to show them the might of imperial military power and thus impress upon them the empire's political objectives. For example, after the fall of Hunza in 1891, Dunmore met the rulers of Hunza and other tribal states in Rawalpindi,

a military garrison town in British India. He writes, "The Government had sent them [the tribal leaders from Hunza and Nagar] down into India to give them some sort of idea of England's power in that country, with a view to their returning to their native states and informing the hill tribes how absolutely futile it would be on their parts to ever attempt to measure strength with such a power as that of Great Britain" (1894, 3).

25. As well, his reference to Alexander played on popular contemporary European theories about the link between the Macedonian conqueror and certain northern Indian tribes that even today have conspicuously "Aryan" features and pale skin color. The popularized version of this theory can be seen in Rudyard Kipling's (1899) *The Man Who Would Be King*.

Chapter Three. Frontier Matters

1. By the end of the nineteenth century, the British had developed an elaborate system of governing the native states of India. Some of these states were fully incorporated into the imperial fold, others were brought under a direct tributary relationship, and yet others were made to acknowledge the supremacy of British power without actually surrendering their sovereignty (Lamb 1968; Robb 1997; Ramusack 2004). The princely state of Kashmir was an important frontier state that theoretically enjoyed autonomous status as far as its system of internal governance was concerned, but its external affairs were a different matter.

2. Chad Haines has argued that British frontier policies between 1850 and 1950 successfully reoriented Hunza's relationship southward, as opposed to its previous northward orientation (2004, 536). This chapter supports Haines's assertion and argues that British policies in the first half of the twentieth century were a turning point in the political history of Hunza.

3. In 1898, Hunza sent fifteen ounces of gold as tribute to China in return for the cultivation of Raskam. The British government in return received 1,070 rupees' worth of gifts (Macpherson to Cobb, April 10, 1914, R/2/1082/265).

4. Extract from the *North China Herald* dated the May 22, 1891; abstract of *Peking Gazette*; tribute offering from Kanjut, R/2/1062/16.

5. Précis of papers relating to rights of the Kanjutis in the Raskam valley, November 16, 1903, R/2/1080/256.

6. Ibid.

7. T. C. Pears to deputy secretary of state, GoI, September 26, 1905, R/2/1080/257.

8. Curzon to secretary of state, GoI, January 26, 1905, R/2/1080/257.

9. Ibid.

10. Ibid.

11. A. D. Macpherson to H. V. Cobb, April 10, 1914, R/2/1082/265.

12. Précis of correspondence regarding the Raskam question, n.d., R/2/1082/265.

13. Ibid.

14. H. V. Cobb to secretary of state, GoI, May 12, 1914, R/2/1082/265.

15. A. D. Macpherson to H. V. Cobb, March 14, 1915, R/2/1082/266.

16. H. V. Cobb to secretary of state, GoI, May 12, 1914, R/2/1082/265.

17. H. V. Cobb to P. Zachariah, foreign secretary, GoI, May 16, 1914, R/2/1082/265.

18. Hsu Kuo-Chen, taoyin of Kashghar, to mir of Hunza, October 21, 1922, L/P&S/12/3292.

19. Taoyin of Kashghar to mir of Hunza, September 13, 1923, R/2/1086/312.

20. Political agent in Gilgit to resident in Kashmir, September 20, 1923, R/2/1086/312.

21. Political agent in Gilgit to GoI Foreign and Political Department, September 11, 1937, R/2/1085/296.

22. Political agent in Gilgit to GoI Foreign and Political Department, September 18, 1937, R/2/1085/296.

23. Political agent in Gilgit to GoI, December 29, 1938, R/2/1085/296.

24. Under the Madhopur Treaty of 1872 between the Kashmir state and the British, which led to the establishment of the Gilgit Agency, the British political agent was responsible only for the defense of the frontier (Gerard 1897). But the British were concerned that leaving the internal matters of the agency under the rule of the Kashmir state would lead to unrest in the region.

25. Francis Yunghusband to secretary of state, GoI (confidential), June 29, 1908, R/1 /1/359.

26. Resident of Kashmir to director, Frontier Circle, Simla, Survey of India, November 11, 1927, R/2/1067/93.

27. Ibid.

28. Ibid.

29. Prime minister of Kashmir to resident in Kashmir, March 3, 1942, R/2/1070/137.

30. Political agent in Gilgit to assistant to the resident in Kashmir, September 8, 1936, L/P&S/12/3299, f. 29.

31. B. E. M Gurdon to political agent in Gigit, May 23, 1894, R/2/1079/251.

32. Political agent in Gilgit to resident in Kashmir, September 13, 1894, R/2/1079/251.

33. Ibid.

34. He describes Curzon's travel to the Pamir through Hunza, describing Curzon as thoroughly inept at walking in the mountainous terrain. At one point Curzon had to be carried on the back of Hunza men to cross the Khunjerab River. Of course, Curzon does not mention any such events in his own writings (Nazeem 2001).

35. This practice is well described by Bayly (1999), who argues that contrary to the dominant theory of British-Indian interaction, in which little room is given to mutual adoption of techniques of governance and ruling styles, the British built on the existing Mughal political and cultural structures to claim legitimacy for their rule. For a popular elaboration on this concept, see William Dalrymple, who argues that many instances now show that the "steely dualism of the Empire had been broken" (2004, xlv).

36. Resident of Kashmir to secretary of state, GoI, October, 22, 1938, f. 38/4, R/2/1086/301.

37. Ibid.

38. Ibid.

39. About the British intermingling with the royal family, Schomberg wrote, "A Gushpur is a member of one of the ruling families who enjoys, owing to his social status (blood, kinship), a certain prestige with, unhappily, certain privileges: he belongs in fact to a decadent, arrogant, greedy, privileged class, which will not work but expects to be supported in idleness. . . . Every European is pestered by these hungry drones. It is impossible not to feel sorry for them, for they are largely the victims of tradition and environment, but it is equally impossible to help them. They intrigue and cabal and, in their way, are a nuisance. In the good old days war, sudden death, and other eventualities, incidental to a wild and undisciplined existence, modified the problem. Now, alas! It increases with the blessings of peace" (1935, 22).

40. For a complete analysis of the internal political dynamics of the Hunza state, see Stellrecht 2006.

41. E. N. Cobb to first assistant to resident in Kashmir, August 14, 1945, R/2/1071/162.

42. As Emily Lorimer wrote: "Everything the Hunzakuts do is beautifully done, and their methods are in greatest contrast to the slovenliness of Nagir; you notice this everywhere, down to the tiniest detail; it may be merely because water is scarcer and life harder that their fields are more scrupulously level, their walls more ingeniously perfect, their cut swathes more exactly aligned, but I incline to think that the cause lies deeper; in the difference of race and temperament" (1939, 72).

43. Leila Blackwell reported a similar experience of staying at Ye Olde Pigge Whistle at Bunji (1950, 23).

Chapter Four. Rural Romance and Refuge from Civilization

1. This discourse would later come to be known as organic farming, as described by Sir Albert Howard and Lady Eve Balfour (Heckman 2006).

2. In an attempt to prove the superiority of the northern diet, he fed different groups of rats a range of diets representing different areas of India for 140 days, a period equivalent to twelve years of human life.

3. In a later experiment, McCarrison fed the northern Indian diet to one group of rats and the diet of the "poorer classes in England," comprising white bread, margarine, sweet tea, boiled cabbage and boiled potato, and tinned meat and tinned jam "of the cheaper sorts." Again, he found that the first group flourished, both physically and socially, while the rats in the second group were unhealthy and fought among themselves (1961, 29).

4. It had been established in medical science that diseases were often caused by lack of vitamins, a view with which McCarrison entirely agreed, but he argued that the medical procedure of narrowing the causes of individual diseases down to particular vitamin deficiencies was a fragmentary approach to the overall problem (Aykroyd 1960, 416, citing McCarrison 1937, 1945). McCarrison called instead for a holistic approach in which overall nutrition was at the center, and in which it was not only important what food was consumed but also how it was produced.

5. The Indian Agriculture Service came into being in 1906, a result of the need for improvement in agricultural productivity and resistance to famines. Lord Curzon, who was the viceroy of India at that time (1899–1905), was the political will behind introducing such research and experimentation through science in Indian agriculture (Arnold 2000, 151). Two years after the establishment of the Pusa station, Albert Howard pioneered the idea of composting—returning to the soil its basic components to maintain its healthy structure.

6. This book was central in providing the foundations for the establishment of the Soil Association, the United Kingdom's leading organic organization today. The Soil Foundation's founder, Lady Eve Balfour, met with Wrench and McCarrison before starting the organization in 1946. See http://homepages.tesco.net/~Haughley/soilass.htm. Lady Balfour wrote her own book in 1943, *The Living Soil*, which "presented the case for an alternative, sustainable approach to agriculture that has since become known as organic farming" (http://www.soilassociation.org/web/sa/saweb.nsf/Aboutus/History.html).

7. In the last chapter, we saw that Younghusband ascribed the mental fitness of Hunza's Wazir Dadoo to the natural conditions of Hunza, and he even condoned caravan raiding as part of the natural fitness of the people's physique. That was a different romantic discourse from the one that we are observing here. The earlier romantic discourse celebrated nature; this discourse celebrates culture.

8. Rodale is now a global media company that publishes a number of the world's best-selling "health and wellness" magazines (see www.rodale.com).

9. This is similar to the Swiss village described by Robert Netting as "an island in the sky" (Netting 1981) with a perfect balance between humans and their environment, untouched by the changes that had taken place elsewhere. Netting later revised his analysis of the closed ecological system of the Swiss village of Torbel and confessed that he had been "guilty of ecosystemic fallacy" (1990, 229).

10. Jean Shor also wrote an article with her husband when she traveled to Hunza for the *National Geographic Magazine* in 1953. The title of the article was "At World's End in Hunza."

11. Contemporary research on Hunza tends to support the work of the Japanese doctors, showing that fear of famine and prevalence of disease were, in fact, common features of village life in the mid-twentieth century (Halvorson 2003). Nigel Allan argues that by fixing tax in the form of wheat, the state of Hunza had actually contributed to the poor health of Hunza. Allan argues that from the point of view of nutritional value, it would have been much more beneficial if vegetables had been grown instead (1990, 412).

12. For example, Leaf frankly accepts that his findings are based on "impressions" rather than actual verification of claims about long ages. "In Hunza the dating problem was particularly difficult. There is no written form so no record exists. In some instances, however, the Mir/ruler of Hunza could, from personal history, verify ages. In short, I was unable to confirm exact ages in Hunza yet I had the definite *impression* of an unusual number of very vigorous old folk clambering over the steep slopes that make up this mountain land" (1973, 96; emphasis added).

13. In official classification, a das would be identified as "wasteland," but for the local people it usually represented uncultivated land, mainly used for winter grazing of sheep and goats but also having potential for growing crops.

14. Mir Jamal Khan was the mir of Hunza from 1946 to 1974. I will from now on simply refer to him as "the mir" rather than writing his full name.

15. For lack of a better word, I am using the term *manage*, but it could also be understood as performance, à la Ferguson (1999), Tsing (1993), and even Goffman (1959), who describe the way in which marginal players in particular put on performances to satisfy those with whom they are interacting.

16. An interesting feature of this period is the emergence of the cold war. Indeed, we see some visitors to Hunza during this period explicitly pointing toward the development of a possible flashpoint in this region. Ian Stephens, an English traveler, went on a journey in northern Pakistan in the early 1950s just when the traffic of foreign travelers and researchers on Hunza health began to increase. This was a time when Communist China was emerging as a major power in the East. Stephens's "Introduction to the American Edition" opens: "This book is about Muslim countries—countries on the far side of the globe in which America until recently has taken little interest. . . . The book concentrates on the northerly tract of West Pakistan and the neighboring lands to the East and West, a restless part of the world always, where many of mankind's great affairs have been decided, and races and rulers have been made or smashed; where during the last ten years much bloody commotion has occurred owing to the break-up of British rule in India, and where now, threatening to move into the resulting partial power-vacuum, the forces of Russian and Chinese Communism stand close. Against them there is not much defense. The armies of the new India and Pakistan, such as they are, still point against each other, rather than towards the Communist North" (1955, foreword).

Chapter Five. The Origin of a Nation

Parts of this chapter's section "Hunza in the Pakistan Nationalist Discourse" originally appeared in *Tsantsa*.

1. I used the word *Culture* with a capital C and in quotation marks to denote the popular meaning of the word as understood in the media and in official languages. This meaning is associated with material culture and its representation purposefully made for display, such as in museums.

2. I will discuss the issue of tourism in detail in chapter 7.

3. This development affirms Robert Chambers's (1983) observation that development (and conservation) projects follow road networks.

4. This policy is tied to the concern of the Pakistani state that if Hunza is formally integrated into the Pakistani nation-state, it will imply that the state is conceding its claims over other parts of Kashmir still under Indian control.

5. Named Khyber-Pakhtunkhwa in 2010.

6. Unlike, say, in Indonesia, where nation and nationalism are tied through national language and state-led development, Pakistan does not see Hunza as a project for ethnic assimilation or development. Development from the Pakistani state perspective is a loose and malleable term (Errington 1998, 419).

7. As we know, during the 1980s, the United States and Saudi Arabia led the cultural indoctrination of radical Islamists in Pakistan as part of a strategic investment in developing resistance to the Soviets. Zia used this strategy to mount insurgency in Indian-held Kashmir.

8. We notice some parallels between the official Pakistani nationalist discourse on origins of nations and the British Orientalist discourse on the origins of the Aryan race and a golden civilization. Both rework Hunza into a remote and distant region, both in physical geography and temporally.

9. Toor also discusses the adoption by the Pakistan People's Party of a report written earlier by Faiz Ahmed Faiz, the iconic leftist progressive poet, outlining a cultural policy that could accommodate Pakistan's ethnic and cultural diversity (2005, 331).

10. As Crossley, Siu, and Sutton argue, "Nations have, as a historical necessity, only one 'national' group. Indeed all modern national republics can be shown to have defined their national populations through the backward process of identifying their 'ethnic' groups" (2006, 3).

11. As Pakistani political commentator Nadeem Paracha (2013) writes, "Indigenous Pakistani folk culture and music were aggressively patronised by the populist government of Z. A. Bhutto. Some analysts suggest that this was at least one part of his regime's strategy to co-opt nationalist sentiments simmering among Sindhi, Baloch and Pushtun nationalists."

12. It does not mean that Bhutto abandoned the Islamic slogan; rather, he used it in its more pluralistic and universal form, especially to attract the rural population.

13. The Gilgit Agency was combined with Baltistan, a former district of the Kashmir state, which also broke away from it in 1947 and came under Pakistani administration to form the Gilgit-Baltistan of Pakistan. This move was welcomed by the majority of the people of the Gilgit Agency and the states of Hunza and Nagar. A small portion of the region of the Kashmir state that was not part of the Gilgit Agency but came under Pakistani administration is known as Azad Kashmir. Pakistan has maintained a different position to the British policy of separating Gilgit-Baltistan (Gilgit, Hunza, and Baltistan) from the actual state of Kashmir.

14. An additional source of appreciation of Hunza and its distinct "mountain culture" within the Pakistani state is derived from Westerners' attraction to this region. Hunza is a popular tourist destination in the diplomatic and foreign community in Islamabad. To many Westerners, especially western Europeans and Americans, it is the one place in Pakistan that is most unlike Pakistan. Today, one of the main markers of a Pakistani ethnic society's rating criteria is its level of "fundamentalism." The people of Hunza are considered relatively liberal, tolerant, and nonviolent, which is an accurate representation of them, among most middle-class and upper-class Pakistanis. Hunza in this way is connected, in its outlook toward life, to the minority liberal and educated classes of Pakistan. Because of their relatively apolitical stance, managed very astutely and highly politically, the people of Hunza remain outside the current poisonous national discourse of who is a real Muslim and hence eligible for Pakistani citizenship. Their Ismaili identity as Muslims in Pakistan is precariously maintained, subject to scrutiny and rejection, as is happening in the case of the Shias of the region.

15. During the reign of the Persian Qajjar dynasty in the eighteenth century, the ceremonial title of "Aga Khan" was given to the Ismaili imam of the time and has continued to the present date. In the 1840s, the first Aga Khan moved to Bombay (Daftary 1990), where the community established itself, and in the early twentieth century, Aga Khan III, Sir Sultan Mohammad Shah, played a major role in financing the Pakistani movement. He was also the president of the All India Muslim League between 1906 and 1913 (Aziz 1998). The present imam, Prince Karim Aga Khan, lives in Aiglemont outside Paris. In addition to his role as spiritual leader of the Ismailis, he is the head of a number of philanthropic institutions under the umbrella of the AKF. Today's Ismailis in Hunza were converted to Ismailism from Twelver Shi'ism in the sixteenth century by *da'i* (Ismaili religious missionaries) sent from Central Asia.

16. Print capitalism is a concept forwarded by Benedict Anderson (1991), who stated that nations emerge as imagined communities by sharing a common language and discourse generated by the printing presses of capitalist societies.

17. Some local people in Hunza, particularly those who had been active in the struggle during the 1960s and 1970s to abolish the mirdom and achieve provincial status for the Gilgit-Baltistan, argue that although the AKRSP has brought welcome economic development to the area, it has had a negative impact politically. Ghazi Mohammad, the hotel owner who was actively involved in politics in Karachi in the 1960s and who had told me that the local council structure had a depoliticizing effect, made similar charges against the AKRSP. He complained to me that many of his radical colleagues and friends from those days ended up working for the AKRSP and other local AKF institutions, becoming depoliticized not only in their actions but also in their thinking. Thus, according to Ghazi Mohammad and others I spoke with, the process of political maturation and consciousness has actually been stymied by the introduction of "development" into the area. For Ghazi Mohammad the AKRSP literally represented the Fergusonian "Anti-Politics Machine" (1990). As Ghazi Mohammad told me, the goal of full political rights, of being seen as equal to other provinces in the nation, a goal that is not only a matter of material gains but also of dignity and self-respect, has been overshadowed by the short-term gains of infrastructural improvement and economic development.

18. Shimshal has a historical reputation that the mir used to send criminals and political prisoners there because of its remote and isolated location.

Chapter Six. On the Edge of the World

1. Perhaps the most effective definition of a transhumance economic system is given by John Myers in his 1942 essay "Nomadism." "In transhumance the community has a fixed abode, and may be maintained, to some extent, by some form of plant cultivation; but the herds have distinct seasonal pastures, and are transferred from higher to lower, and back again. Either the higher or the lower pasture, and usually the lower, is adjacent to the settlement and the fields; but sometimes there is a secondary village, or scattered summer huts, on the upper, and often a crop of hay is taken here before the cattle are admitted to graze; this serves for winter fodder, to supplement or replace winter grazing" (16–17).

2. This is the same area that once had significance in the context of the Great Game in the late nineteenth and early twentieth centuries.

3. Anthropological literature on the role of women in pastoralism shows that control of animals and women goes hand in hand because they represent the two most important sources of power: economic resources and human labor. Evans-Pritchard's (1940) insight that pastoralism is a male-dominated activity in which men's control of livestock allows them to control women and their reproductive capacity and labor, mainly through the institution of livestock as bride wealth, is a case in point. For men to control livestock and women, they must prohibit women from the pastoral economy and activities. This point is further elaborated by widespread studies that show women are considered ritually impure and thus are forbidden to herd in many pastoral societies (Schneider 1980). Peter Parkes's (1987) study of the Kalash transhumance economy shows that it is the men who control the livestock, and women's presence near the livestock herd is considered taboo, creating a sharp opposition and antagonism between sexes. Parkes argues that such a worldview is common in the Mediterranean, where the natural environment is divided into the pure realm of herding by men and the impure realm of the women in the village (638). Other studies from the region show that either because of the strict Islamic tradition of purdah (Glatzer and Casimir 1983), or because of socioeconomic pressures (Goldstein and Beall 1989), women are forbidden from going outdoors for herding practices although they may engage in milking and shearing livestock. Shimshal, then, represents an exceptional case in which the role of women in herding yak is not only appreciated but encouraged.

4. Indeed, I have seen many Punjabi and other Pakistani tourists from "downcountry" mistake them for foreign trekkers and initiate conversation in English.

5. Pamer is referred by the Shimshali to the undulating alpine plains between twelve and seventeen thousand feet surrounding Shimshal Pass. This should not be confused with Pamir, the mountain range and geographical region in Pakistan Afghanistan and Tajikistan, which is located northwest of Pamer. The similarity in names is due to the same kind of landscape, that is, high tableland, in both regions.

6. This is the term used to refer to small power-generation systems that harness energy from small streams.

7. I will discuss the design of the house in detail in the next chapter.

8. The danger of lake burst had posed a substantial threat not only to Shimshal but also to communities downriver. There existed a unique system of warning whereby the people down the valley were informed about the bursting of the lake. In times of danger, people were deployed at various viewpoints on mountaintops all the way to Baltit. When the lake started to spill over the glacier, the persons on duty close to the lake would make a fire that would be visible from the next viewpoint. The person there would then light a fire visible from the next, and so on. In this way the message was conveyed to Hunza in a few minutes. This reminds me of *The Lord of the Rings*, in which Gondor summons the help of Rohan against the armies of Sauron using the same system.

9. Indeed, this was the case in a legal sense as well. As we shall see, the boundary of the Khunjerab National Park started exactly where the village ended.

10. According to the version produced by Butz (1992, 7–11), this man fled Hunza during the political crises that led to the breakup of the Nagar and Hunza states. He came to Sarikol looking for a wife.

11. According to Butz, they fled the country because some Sarikoli men became jealous of Mamu Shah's flock and planned to murder him. Khadija heard the plan and they both fled Sarikol (1996, 8).

12. It is interesting to note the implication of this, which is that this place was settled before. I tried to find out who could have been there before Mamu Shah's arrival, but Shimshalis claim not to know.

13. The term *saint* in the Pakistani context means a living holy person with miraculous God-given powers.

14. A stage is officially described as the distance a porter should cover in a day for a set fee, a definition that leads to endless tensions between outside tourists and local porters. The former, usually, argue that whatever distance is actually covered should count as a single stage, regardless of any predetermined idea of what distance constitutes a stage. Porters, however, insist—correctly, I am sure—that their daily wages should not be determined by the ground covered in a day but by the total length of the distance covered. Thus, if more than one stage is covered in one day, which often happens, as two stages are usually about three to four hours apart, then they should be paid by the stage, not by the day. This formulation of wages is based on the fact that distance alone does not capture the difficulty of the terrain covered.

15. Pareeans are usually made along the path to the Pamer, where there are only a limited number of such sections, and a family after whose deceased relative a particular section has been named has the exclusive right to that section's future improvement and maintenance. For example, there are a total of nineteen pareeans between Shimshal and Pamer settlement, and there are 120 families in Shimshal. If another family wants to dedicate improvement work to a deceased relative on a particular section that has already been named after someone else's deceased relative, that family has to first seek permission from its current name holder. The entitlement to maintain a pareean is then traded on a permanent basis.

16. There are two paths the lead up to Pamer from Shimshal village.

17. This path was one set aside exclusively for human use or smaller livestock. The yaks took a different, longer route.

18. The argument was originally made by Myers in his *Man* article in 1942.

19. *Kooch* is a Persian word that means "departure." In the context of the Shimshali transhumance economy, it means the seasonal migration of animals and people to and from the pastures. So there is a summer kooch, when Shimshali women leave with goats and sheep in mid-May for the Shimshal Pass to spend the summer, and then there is the return kooch in October when they return with them to the village.

20. The term *mergich*, the realm, and *mergichon* are polysemous. *Mergichon* could also mean spirits of holy people such as *Awliyas*, saints with miraculous powers. So their help is sought not only when on pasture but also when one embarks on a journey or a difficult task. *Mergich*, in addition to being a realm, a spatial concept, can also mean a duration or period and is hence a temporal concept. So the one-week period after

which women arrive on the pasture and during which they don't consume fresh milk is also called the period of mergich.

21. The mergichon can also be a girlfriend of a shpun, this alerting us to masculine desires in a separated realm of the female world.

22. Descola reports a similar belief among the Achuars of the Andes, who believe that wild animals belong to the spirit of the forest (2013, 41). Descola claims that the Achuars do not make strict distinction between the wild and the domesticated; the Shimshalis, however, actively mark this distinction.

23. Marcel Mauss (1979) has discussed similar social organization among the Eskimos, whose "social morphology" varies with seasonal variations based on the dispersal and coming together of the band.

24. Shahrani (1979) describes a similar practice among the Kirghiz nomads of the Pamir region, where the surplus animals were redistributed among poorer members of the society. He disagrees with Spooner's (1973) claim that rich farmers often sell their surplus animals to buy immovable property and ultimately become sedentary. The situation in Shimshal today is more in line with Spooner's observation.

Chapter Seven. Strange Strangers in the Land of Paradise

1. Writing about a small isolated community of Toraja in Sulawesi, Indonesia, Kathleen Adams states that such representations by the tourism industry often overlook other features of a community such as hierarchical social organization and orientations toward money and land. She states that these outsider representations of Torajas' otherness have been manipulated by the touristically enculturated Toraja; "Local communities are hardly passive, and touristic phenomena are actively manipulated by local community members for their own objectives" (2005, 47). Millie Creighton, writing about the marketing of *furusato*, or "hometown," by Japanese travel companies to urban Japanese, identifies the same dynamics. "In the pursuit of nostalgia, furusato has been decontextualized. It is frequently portrayed by images of rustic landscapes, dilapidated shrines, and remote anonymous train stations. With the mass marketing of furusato, specific place identity is masked, so that any rural location may symbolically be experienced as anyone's furusato—even for those who grew up in cities" (1997, 239).

2. I am referring here to the transhumance migration cycle in which women returning from the summer pastures and men returning from the winter pastures, who have temporarily become strangers to the villagers, are reintegrated into the village and domestic life.

3. Though not quite as isolated as Shimshal, Hushe is a small village located at the furthermost end of a valley in Baltistan; it is clearly not only its mountainous vistas that make it comparable to Shimshal but also its isolation from outside influences.

4. This number increased to eighteen thousand in 2012.

5. Ironically, it is the company based in Nepal that notes Shimshal is a Wakhi village, whereas the Pakistani companies represent it as part of Hunza culture.

6. There are, of course, regional variations between what is generally called a Wakhi house in the Wakhi villages in Hunza and in neighboring countries (Afghanistan, Tajikistan, and China). But within Pakistan, as I have observed, these are minor and do not alter the basic "superstructure" of the house.

Chapter Eight. Romanticism, Environmentalism, and Articulation of an Ecological Identity

1. Other nonconsumptive uses include research and education.

2. As William Cronan stated about wilderness, "Far from being one place on earth that stands apart from humanity, it is quite profoundly a human creation" (1995, 69).

3. But intellectually wilderness was understood as many different things and was appreciated for variegated reasons. One central meaning attached to the idea of wilderness was its removal from mainstream society and civilization and its position as an escape from them.

4. As Donald Worster (2006) shows, the early conception of ecology was driven by utilitarian and Arcadian views of nature, both of which saw nature through an anthropocentric lens.

5. Indeed, wildlife was seen as supporting the Native Americans, so it was also exterminated in an overall effort to wipe out the Native Americans.

6. In the case of the KNP, Schaller wanted to save the Marco Polo sheep but not, say, the ibex or the blue sheep, because the former held a special aura and status in Victorian sportsmen circles and the literature they produced, in which Schaller was well versed; moreover, he worked for an organization, the Wildlife Conservation Society, whose intellectual genealogy could be traced to that era.

7. Schaller wrote, "On my return I wrote a report to the government in which I proposed the establishment of Khunjerab National Park. Dr. Rizvi, who first helped me to obtain permission to visit northern Hunza, now made certain that my report reached Prime Minister Zulfikar Ali Bhutto. The Prime Minister read my notes, agreed with the concept and the proposed borders, and ordered the park established" (1988, 99).

8. For a critique of the role of environmental organizations in the establishment of the KNP, see Knudsen 1999. Are Knudsen argues that neither Schaller nor IUCN used any scientific data to prove that there was actual degradation taking place in the KNP area due to grazing and human impact. Moreover IUCN ignored some of its own policies in the designation and management of the KNP.

9. The community-based conservation approach did not emerge from an anti–protected areas approach, although this was the political context in which it became popular. The seeds of this were found in the early conceptions and legislation about the national parks, according to which some native people were allowed to stay and continue their lives inside the park because they were considered part of nature (Nash 1967; Neumann 2002, 125). Neumann, for example, writes that colonial officers, while enforcing parks regulations in Africa, maintained that "sometimes a 'naïve' presence in the parks may be tolerated" (125).

10. The SNT established a comprehensive park management plan that it presented as a community-based alternative to the government-backed management plan for conservation in the KNP. The Shimshalis claim that rather than having the management of the park in the hands of the directorate of the KNP, it should be managed, including responsibility for finances, by the SNT. SNT has a president, secretary, treasurer, task force, and a board of governors.

11. Undoubtedly, the association of organic, community-oriented society with rural areas and of alienated and disenchanted society with urban areas, the latter a center of a capitalist base, is a continuation of an enduring theme of changing conceptions of nature and culture. It found a new expression in environmental conservation and policy arenas in the 1980s, one in which local communities emerged as the stewards of nature against the onslaught of modern civilization.

12. Despite having opposite material consequences, these two discourses—one advocating nature conservation by removing people from land and the other advocating working with them—existed and continue to exist, side by side, with one never completely able to overcome the other and both states and local communities using them opportunistically.

13. Peter Parkes (2000) notes in the case of Kalash that internationally funded development and minority environmental projects make romantic representations of natural resource management practices of the Kalash, an animistic group in the Chitral, yet at the same time represent such practices as under threat, thus justifying intervention. Such representations also appear in the writings of many researchers who come to do a short stint in the mountains on "research tourism."

14. The effects of such a discourse were documentation, cataloguing, and ordering of local knowledge in an attempt to preserve it before it was lost. This romantic discourse was, then, opposite to Schaller's wilderness romanticism in which the people of Hunza and Shimshal are represented as careless and unsophisticated people whose actions regarding local nature are detrimental and who need the guiding and authoritative hand of a disciplining state.

15. As described earlier, the shpuns are the nine men who graze the village yaks during the winter months near the Chinese border.

16. Of course, these were not the elders of Shimshal; rather, they were the young blood of Shimshal. I use the term *tribal elders* to show the similarity of the role that Tsing's (1999) tribal elders and members of the SNT task force play. Like the tribal elders of the Dayaks of Indonesia, the SNT task force members articulate a particular communal identity in environmental and development discourse.

17. Despite a universal opposition to the KNP, the Shimshalis are not united on one platform against it. Internal tensions, especially those having their origins in shifting tribal and clan alliance and in local and regional politics, are held side by side with the tendency to unify against the park. Most Shimshalis know of these tensions and seldom tried to give me an impression otherwise. The numberdaar of the village, for example, told me candidly that the members of the SNT were corrupt and ineffective. According to him, the members of the SNT Task Force are looking for funds, status, and networking for personal gain and do not really have the concerns of the community

at heart. He said the SNT had deceived the simple Shimshalis by exploiting their cause for their personal benefits, such as trips to foreign conferences and the opportunity to meet people in power and land lucrative jobs.

18. I am using the term *indigenous people* for Shimshal in a very broad sense. Though Pakistan does not have an official or popular category of indigenous people, as countries such as Indonesia, Australia, and others have, the general characteristics of the Shimshalis and their concern for cultural survival make them quite similar to indigenous people.

19. Indeed, as Tsing (1993, 14) states, such claims find their potency in a lack of imagination in modern urban culture, which is unable to find difference within itself—hence, it has to construct a chasm of time and space across which a unique and different culture can exist.

20. Indeed, many scholars have critiqued the movements for indigenous rights for focusing too much on remote and isolated—but exotic—groups while ignoring very common forms of marginality and powerlessness among peasants and nonexotic peoples (Gupta 1998).

21. The only thing that seems to change the legal status of the KNP and hence give the Shimshalis real strategic advantage is the political economy of conservation itself. In the late 1990s, the regional government started a trophy-hunting program in the region surrounding the KNP, in which communities benefited financially by protecting game animals such as the ibex and markhor. Pressure from the local sports outfitter, working in conjunction with local communities, led to the opening of certain portions of the KNP for trophy hunting in order to provide economic benefits to the Shimshalis. This change in practice is likely to bring change in the policy toward management of the KNP by the government of Gilgit-Baltistan.

Epilogue

1. I am not arguing that remoteness as a social category did not exist before the height of the age of European modernization in the mid-nineteenth century, marked by industrialization, urbanization, commodity production, and the spread of colonialism and markets. Many scholars have shown how the Greeks (Regar 2002) and Romans (Freedman 2008) imagined distant places and markets and what kinds of conceptions they held about them. However, I suggest that remoteness has a particularly enduring association with modernity.

2. He wrote, "The very shaping of history now outpaces the ability of men to orient themselves in accordance with cherished values. And which values? Even when they do not panic, men often sense that older ways of feeling and thinking have collapsed and that newer beginnings are ambiguous to the point of moral statis. Is it any wonder that ordinary men feel they cannot cope with the larger worlds with which they are so suddenly confronted? That they cannot understand the meaning of their epoch for their own lives? That—in defense of selfhood—they become morally insensible, trying to remain altogether private men? Is it any wonder that they come to be possessed by a sense of trap?" (2000, 4).

3. For Mills (2000), sociological imagination is a positive mental attribute geared toward some action. It is a methodology that explains facts better because one is looking at facts from an alternative perspective. For Mills, then, it is a reflexive approach to understanding social reality.

4. Here I mean that it is generally in faraway and distant places, rather than in near and proximate ones, that the effects of modernization are studied. Hence, these faraway, remote places are universalized as "local" places in development and academic literature.

5. At the global level, a nuanced understanding of northern Pakistan remains sketchy, so the entire mountainous northern region is often termed as the hotbed of al-Qaeda and the Taliban. A recent best seller in the United States, *Three Cups of Tea*, about the humanitarian work of an American, painted Baltistan and parts of Hunza as the backyard of the Taliban, when the fact is that the Taliban is a mainly Pashtun and Punjabi movement. Thankfully, the inaccuracies of the book were exposed in *Three Cups of Deceit*, a book by John Krakaur, and in a piece by CBS's *60 Minutes*.

6. This fact that remoteness no longer provides as solid a tactical advantage to insurgents and rebels as it did for the last 150 years may be one reason why Osama bin Laden was found not in the remote and rugged terrain of the Pakistan-Afghanistan borderlands but rather in the city of Abbotabad, deep in the heart of Pakistan and close to its capital city.

References

Adams, K. 2005. Generating theory, tourism and "world heritage" in Indonesia: Ethical quandaries for anthropologists in an era of tourist mania. *Napa Bulletin* 23:45–59.

Aga Khan Foundation. n.d. http://www.akdn.org/rural_development/about.asp (accessed June 30, 2014).

Aga Khan Rural Support Program. 1993. *Annual review*. Gilgit: Aga Khan Rural Support Program.

Alder, G. J. 1963. *British India's northern frontier*. London: Longmans, Green.

Ali, Farman. 2007. Constitutional package for Northern Areas. *Daily Dawn*, October 24. http://www.dawn.com/news/272573/constitutional-package-for-n-areas (accessed July 7, 2014).

Ali, S. S., and J. Rehman. 2001. *Indigenous peoples and ethnic minorities in Pakistan: Constitutional and legal perspectives*. Nordic Institute of Asian Studies, monograph series, no. 84. Richmond, UK: Curzon.

Allan, N. 1986. Accessibility and altitudinal zonation models of mountains. *Mountain Research and Development* 6 (3): 185–94.

———. 1990. Household food supply in Hunza Valley, Pakistan. *Geographical Review* 80 (4): 399–415.

———. 1993. Dardistan. In *Encyclopædia Iranica*, 7:26–27. New York: Routledge and Kegan Paul.

Andereck, K. L. 2005. Evaluation of a tourist brochure. *Journal of Travel and Tourism Marketing* 18 (2): 1–13.

Anderson, B. 1991. *Imagined communities*. Princeton: Princeton University Press.

App, U. 2009. *William Jones's ancient theology*. Sino-Platonic papers, no. 191. Edited by V. H. Mair. Philadelphia: University of Pennsylvania, Department of East Asian Languages and Civilizations.

Appadurai, A. 1988. Putting hierarchy in its place. *Cultural Anthropology* 3 (1): 36–49.
———. 1996. Here and now. In *Modernity at large: Cultural dimensions of globalization*, edited by A. Appadurai, 1–26. Minneapolis: University of Minnesota Press.

Ardener, E. 1989. Remote areas: A theoretical consideration. In *The voice of prophecy and other essays by Edwin Ardener*, edited by M. Chapman, 211–23. New York: Blackwell.

Argent, N. M., and F. Rolley. 2000. Financial exclusion in rural and remote New South Wales, Australia: A geography of bank branch rationalisation, 1981–98. *Australian Geographical Studies* 38 (2): 182–203.

Arnold, D. 2000. *Science, technology and medicine in colonial India.* The new Cambridge history of India. Cambridge: Cambridge University Press.

Aykroyd, R. W. 1960. Obituary notice: Major-General Robert McCarrison. *British Journal of Nutrition* 14:413–18.

Aziz, K. K. 1998. *Aga Khan III: Selected speeches and writings of Sir Sultan Mohammad Shah.* London: Kegan Paul International.

Balfour, E. 1943. *The living soil: Evidence of the importance to human health of soil vitality, with special reference to post-war planning.* London: Faber.

Banik, A., and R. Taylor. 1960. *Hunza land: The fabulous health and youth wonderland of the world.* California: Whitehorn.

Barrow, J. 1831. Prospectus of the Royal Geographical Society (at the first meeting of the Royal Geographical Society). *Journal of the Royal Geographical Society* 1:vii–xii.

Basham, A. L. 1959. *The wonder that was India.* New York: Grove.

Bauman, Z. 1990. Modernity and ambivalence. In *Global culture: Nationalism, globalization and modernity*, edited by M. Featherstone, 143–69. London: Sage.

Bayly, C. 1999. *Empire and information: Intelligence gathering and social communication in India, 1780–1870.* Cambridge studies in Indian history and society. Cambridge: Cambridge University Press.

Beg, U. Q. 1935. *Tarekh-e-Ahd-e-Ateeq-e-Hunza.* Rawalpindi: S. T. Press.
———. 1967. *Tareekh Tameer Central Jamat Khana Gilgit.* Baltit, Hunza, n.p.

Biddulph, C. E. 1891. Russian Central Asia. *Journal of Royal Asiatic Society* 23 (4): 563–97.

Biddulph, J. 1880. *The tribes of Hindoo Koosh.* Calcutta: Office of the Superintendent of Government Printing.

Blacker, L. V. S. 1922. Wars and travels in Turkistan, 1918–1919–1920. *Journal of the Royal Central Asian Society* 9 (1): 4–20.

Blackwell, L. 1950. *Memoir: A year in the Gilgit Agency.* Blackwell Papers. Cambridge: University of Cambridge, Center for South Asian Studies.

Brockington, D., R. Duffy, and J. Igoe. 2010. *Nature unbound: Conservation, capitalism and the future of protected areas.* London: Earthscan.

Brosius, P. 1997. Endangered forest, endangered people: Environmentalist representation of indigenous knowledge. *Human Ecology* 25 (1): 47–69.

Browdin, P. 2003. Marginality and subjectivity in the Haitian diaspora. *Anthropological Quarterly* 76 (33): 383–410.

Brown, W. A. 1998. *The Gilgit rebellion*. Gilgit: Ibex.

Bryant, E. 2001. *The quest for the origin of Vedic culture*. Oxford: Oxford University Press.

Butz, D. 1992. True stories, partial stories: A century of interpreting Shimshal from the outside. Unpublished manuscript, Brock University, St. Catherines, Ontario.

———. 1996. Sustaining indigenous communities: Symbolic and instrumental dimensions of pastoral resource use in Shimshal, northern Pakistan. *Canadian Geographer* 40 (1): 36–53.

———. 2000. Sustainable tourism and everyday life in Shimshal, Pakistan. *Tourism Recreation Research* 27 (3): 53–65.

———. 2011. Narratives of accessibility and social change in Shimshal, northern Pakistan. *Mountain Research and Development* 31 (1): 27–34.

Chambers, R. 1983. *Rural development: Putting the last first*. London: Longmans.

Chohan, S. A. 1985. *The Gilgit Agency, 1877–1935*. New Delhi: Atlantic.

Clark, G. E. 1977. Who were the Dards? *Kailash* 5 (4): 323–56.

Cockerill, G. 1922. Byways of Hunza and Nagar. *Geographical Journal* 60:97–112.

Cohn, B. 1996. *Colonialism and its forms of knowledge*. Princeton: Princeton University Press.

———. 2004. *An anthropologist among the historians and other essays*. New Delhi: Oxford University Press.

Concordia Expeditions. n.d. http://www.concordiaexpeditions.com/shimshal_trip. htm (accessed February 13, 2008).

Conklin, B., and L. Graham. 1995. The shifting middle ground: Amazonian Indians and eco-politics. *American Anthropologists* 97 (4): 695–710.

Creighton, M. 1997. Consuming rural Japan: The marketing of tradition and nostalgia in the Japanese travel industry. *Ethnology* 36 (3): 239–54.

Cronan, W. 1995. The trouble with wilderness; or, getting back to the wrong nature. In *Uncommon grounds*, edited by W. Cronon, 69–90. New York: Norton.

Crossley P., H. Siu, and D. Sutton. 2006. *Empires at the margins: Culture, ethnicity, and frontier in early modern China*. Berkeley: University of California Press.

Csáji, L. K. 2011. Flying with the vanishing fairies: Typology of the shamanistic traditions of the Hunza. *Anthropology of Consciousness* 22 (2): 159–87.

Cullen, B., and M. Pretes. 2000. The meaning of marginality: Interpretation and perception in social sciences. *Social Sciences Journal* 37 (2): 215–29.

Curzon, G. N. 1896. The Pamirs and the source of the Oxus. *Geographical Journal* 8 (1): 15–54.

———. 1908. *Frontiers*. Oxford: Clarendon.

———. 1926. *Leaves from a viceroy's note-book and other papers*. London: Macmillan.

Da Col, G., and D. Graeber. 2011. Foreword: The return of ethnographic theory. *HAU: Journal of Ethnographic Theory* 1 (1), vi–xxxv.

Daftary, F. 1990. *Ismailis: Their history and doctrines*. Cambridge: Cambridge University Press.

Dalrymple, W. 2004. *White Mughals: Love and betrayal in eighteenth century India*. New York: Penguin Books.

Dani, A. 2001. *History of Northern Areas of Pakistan*. Lahore: Sang-e-Meel.

de Certeau, M. 1986. *Heterologies: Discourse on the other*. Translated by B. Massumi. Minneapolis: University of Minnesota Press.

———. 1988. *The practice of everyday life*. Berkeley: University of California Press.

Dearden, P. 1996. Trekking in Northern Thailand: Impact distribution and evolution over time. In *Uneven development in Thailand*, edited by M. Parnwell, 204–25. London: Avebury.

Descola, P. 2013. *Beyond nature and culture*. Chicago: University of Chicago Press.

Dove, M. 2011. *The banana tree at the gate: A history of marginal people and global markets in Borneo*. New Haven: Yale University Press.

Dove, M., J. Hjorleifur, and M. Aung-thwin. 2011. *Bijdragen tot de Taal- Land- en Volkenkunde* 67 (1): 86–99.

Dowie, M. 2009. *Conservation refugees: The hundred-year conflict between global conservation and native people*. Cambridge: MIT Press.

Drew, F. 1875. *The Jummo and Kashmir territories: A geographical account*. London: Edward Stanford.

Duncan, J. 1906. *A summer ride through western Tibet*. London: Smith, Elder.

Dunmore, Earl of. 1893. Journeyings in the Pamirs and Central Asia. *Geographical Journal* 2 (5): 385–98.

———. 1894. *The Pamirs; Being a narrative of a year's expedition on horseback and on foot through Kashmir, western Tibet, Chinese Tartary, and Russian Central Asia*. London: John Murray.

Durand, A. 1900. *The making of a frontier: Five years' experiences and adventures in Gilgit, Hunza, Nagar, Chitral and the eastern Hindu-Kush*. London: Thomas Nelson and Sons.

Edney, M. 1997. *Mapping an empire: The geographical construction of British India, 1765–1843*. Chicago: University of Chicago Press.

Edwards, D. 1989. Mad mullah and Englishman: Discourse in colonial encounter. *Comparative Studies in Society and History* 31 (4): 649–70.

Elias, N. 2000. *The civilizing process: Sociogenetic and psychogenetic investigations*. 2nd ed. Oxford: Blackwell.

Elphinstone, M. 1815. *An account of the kingdom of Caubul: And its dependencies in Persia, Tartary, and India*. London: Longman, Hurst, Rees, Orme, and Brown.

Elsner, J., and J.-P. Rubies. 1999. Introduction to *Voyages and visions: Towards a cultural history of travel*, edited by J. Elsner and J-P. Rubies, 1–57. London: Reaktion Books.

Emerson, R. 1996. Charismatic kingship: A study of state-formation and authority in Baltistan. In *Pakistan: The social sciences' perspective*, edited by A. S. Ahmed, 100–45. Karachi: Oxford University Press.

Errington, J. J. 1998. On the ideology of Indonesian language development: The state of a language of state. *Pragmatics* 2 (3): 417–26.

Esar, F. M. 2001. *Riyasut-e-Hunza: Tareekh aur Saqafut kay Ainey mein*. Gilgit, Northern Areas: Hunny Sara.

Evans-Pritchard, E. 1940. *The Nuer: A description of the models of livelihood and political institutions of a Nilotic People*. Oxford: Oxford University Press.

———. 1950. Social anthropology, past and present. *Man* 50 (198): 118–24.

Fabian, J. 1983. *Time and the other: How anthropology makes its object.* New York: Columbia University Press.

Ferguson, J. 1990. *The anti-politics machine: "Development," depoliticization, and bureaucratic power in Lesotho.* Cambridge: Cambridge University Press.

———. 1999. Rural connection, urban styles: Theorizing cultural dualism. In *Expectations of modernity: Myths and meaning of urban life on the Zambian Copperbelt,* edited by J. Ferguson, 82–122. Berkeley: University of California Press.

Forsyth, D. 1875. *Report of a mission to Yarkand under command of Sir T. D. Forsyth, with historical and geographical information regarding the possessions of the ameer of Yarkand.* Calcutta: Foreign Department Press.

Frank, A. G. 1966. *The under development of development.* Boston: New England Free Press.

Freedman, Paul. 2008. *Out of the East: spice and the medieval imagination.* New Haven: Yale University Press.

French, P. 1995. *Younghusband: The last great imperial adventurer.* London: Flamingo.

Gellner, E. 1983. *Nation and nationalism.* Ithaca: Cornell University Press.

Gerard, M. G. 1897. *Report on the proceedings of the Pamir Boundary Commission.* Calcutta: Office of the Superintendent of Government Printing.

Gibson, C., S. Luckman, and J. Willoughbysmith. 2010. Creativity without borders? Rethinking remoteness and proximity. *Australian Geographer* 41 (1): 25–38.

Giddens, A. 1991. *Modernity and self-identity: Self and society in the late modern age.* Stanford: Stanford University Press.

Gidwani, V., and K. Sivaramakrishnan. 2003. Circular migration and rural cosmopolitanism in India. *Contribution to Indian Sociology* 37 (1–2): 339–67.

Gilmartin, D. 1998. "Partition," Pakistan and South Asian history: In search of a narrative. *Journal of Asian Studies* 57 (4): 1068–95.

Gilmartin, D., and M. Maskiell. 2003. Appropriating the Punjabi folk: Gender and other dichotomies in colonial and post-colonial folk studies. In *Pakistan at the millennium,* edited by C. Kennedy, K. McNeil, C. Ernst, and D. Gilmartin, 97–128. Karachi: Oxford University Press.

Glatzer, B., and M. J. Casimir. 1983. Herds and households among Pashtun pastoral nomads: Limits of growth. *Ethnology* 22 (4): 307–26.

Goffman, Erving. 1959. *The presentation of self in everyday life.* New York: Doubleday.

Goldstein, M. C., and C. M. Beall. 1989. The remote world of Tibet's nomads. *National Geographic,* June, 752–81.

Gordon, T. E. 1876. *The roof of the world: Being a narrative of a journey over the high plateau of Tibet to the Russian frontier and the Oxus sources on Pamir.* Edinburgh: Edmonston and Douglas.

Grove, R. 1996. *Green imperialism: Colonial expansion, tropical island Edens and the origins of environmentalism, 1600–1860.* Cambridge: Cambridge University Press.

Gupta, A. 1998. *Postcolonial developments: Agriculture in the making of modern India.* Durham: Duke University Press.

Haines, C. 2004. Colonial routes: Reorienting the northern frontier of British India. *Ethnohistory* 51 (3): 535–65.

Hall, S. 1992. The West and the rest: Discourses in power. In *Formation of modernity*, edited by S. Hall and B. Gieben, 275–332. Cambridge: Polity/Open University.

Halvorson, J. S. 2003. "Placing" health risks in the Karakoram: Local perceptions of disease, dependency, and social change in northern Pakistan. *Mountain Research and Development* 23 (3): 271–77.

Harms, E., S. Hussain, and S. Schneiderman. 2014. Remote and edgy: New takes on old anthropological themes. *HAU: Journal of Ethnographic Theory* 4 (1): 361–81.

Harrison, S. 2010. China's discreet hold on Pakistan's northern borderlands. *New York Times*, August 26.

Harvey, D. 1999. *The condition of postmodernity*. Malden, MA: Blackwell.

———. 2006. *Spaces of global capitalism: Towards a theory of uneven geographical development*. London: Verso.

Hastrup, K. 1989. Postscript 1: The prophetic condition. In *The voice of prophecy and other essays by Edwin Ardener*, edited by M. Chapman, 224–28. New York: Blackwell.

Hayward, G. 1870. Journey from Leh to Yarkand to Kashghar and exploration of the sources of the Yarkand River. *Journal of the Royal Geographical Society* 40:33–166.

———. 1871. Letters from G. W. Hayward, on his explorations in Gilgit and Yassin. *Journal of the Royal Geographical Society* 41:1–46.

Hearn, J. 2000. *Claiming Scotland: National identity and liberal culture*. Edinburgh: Polygon.

Heckman, J. 2006. A history of organic farming: Transition from Albert Howard's War in the Soil to USDA National Organic Program. *Renewable Agriculture and Food Systems* 21 (3): 143–50.

Herskovitz, M. 1926. The cattle complex in East Africa. *American Anthropologist* 28 (2): 361–88.

Hirtz, F. 2003. It takes modern means to be traditional: On recognizing indigenous cultural communities in the Philippines. *Development and Change* 34 (5): 887–914.

Hoffman, J. 1968. *Hunza: Secrets of the world's healthiest and oldest living people*. New York: Groton.

Hopkirk, P. 1994. *The Great Game*. Tokyo: Kodansha International.

Hughes, D. M. 2008. *From enslavement to environmentalism: Politics on the southern African frontier*. Seattle: University of Washington Press.

Hughes, J. 2013. *Animal kingdoms: Hunting, the environment, and power in Indian princely states*. Cambridge, MA: Harvard University Press.

Hunzai, A. J. 1998. *Hunza kee Lok Kahanian: Sar Zameen-e-Hunza Kee Ujeeb-o-Ghareeb Magar Haqeqee Kahanian* [Folk stories of Hunza: Strange but true stories of Hunza]. Lahore: Sang-e-Meel.

Huskey, L., and T. Morehouse. 1992. Development in remote regions: What do we know? *Arctic* 45 (2): 128–37.

Hussain, S. 2006. Small players in the Great Game: Marginality and representation on the northern frontiers of 19th-century colonial India. *South Asia: Journal of South Asian Studies* 29 (2): 235–53.

Huttenback, R. 1975. The "Great Game" in the Pamirs and the Hindukush: The British conquest of Hunza and Nagar. *Modern Asian Studies* 9 (1): 1–29.

Imanshi, K. 1963. *Personality and health in Hunza valley: Results of the Kyoto University scientific expedition to the Karakoram and Hindukush, 1955.* Vol. 5. Kyoto: Kyoto University, the Committee of the Kyoto University Scientific Expedition to the Karakoram and Hindukush.

Inden, R. 2000. *Imagining India.* Bloomington: Indiana University Press.

Ingold, T. 1986. *Evolution and social life.* Cambridge: Cambridge University Press.

Isserman, M., and S. Weaver. 2008. *Fallen giants: A history of Himalayan mountaineering from the age of empire to the age of extremes.* New Haven: Yale University Press.

Jameson, F. 1988. Cognitive mapping. In *Marxism and the interpretation of culture*, edited by C. Nelson and L. Grossberg, 347–60. Champaign: University of Illinois Press.

Jeal, T. 2012. *Explorers of the Nile: The triumph and tragedy of a great Victorian adventure.* New Haven: Yale University Press.

Karakoram Adventure Holidays. n.d. http://www.karakoramadventureholidays.com/ShimshalTrek.php (accessed February 13, 2008).

Keay, J. 1996. *Explorers of the western Himalayas: The Gilgit game.* London: John Murray.

Knight, F. K. 1900. *Where the three empires meet.* London: Longman, Green.

Knudsen, A. 1999. Conservation and controversy in the Karakoram: Khunjerab National Park, Pakistan. *Journal of Political Ecology* 6:1–29.

Koldys, G. 2002. A political-historical overview of the Kirghiz. In *The Turks*, edited by Hasan Güzel, Cem O uz, and Osman Karatay, 6:224–32. Ankara: Semih Offset.

Kopytoff, I. 1986. The cultural biography of things: Commoditization as process. In *The social life of things*, edited by A. Appadurai, 64–93. Cambridge: Cambridge University Press.

Kreutzmann, H. 1991. The Karakoram Highway: The impact of road construction on mountain societies. *Modern Asian Studies* 25 (4): 711–36.

———. 1993. Challenge and response in the Karakoram: Socioeconomic transformation in Hunza, Northern Areas, Pakistan. *Mountain Research and Development* 15 (1): 19–39.

———. 1995. Globalization, spatial integration, and sustainable development in northern Pakistan. *Mountain Research and Development* 15 (3): 213–27.

———. 1998. The Chitral triangle: Rise and decline of trans-montane Central Asian trade, 1895–1935. *Asien, Afrika Latinamerika* 26:289–327.

Kürti, L. 2001. *The remote borderland: Transylvania in the Hungarian imagination.* Albany: State University of New York Press.

Lamb, A. 1968. *Asian frontiers: Studies in a continuing problem.* New York: Frederick A. Praeger.

Lawrence, J. 1997. *Raj: The making and unmaking of British India*. New York: St. Martin's Griffin.

Leach, E. 1961. *Rethinking anthropology*. London: Athlone.

Leaf, A. 1973. A scientist visits some of the world's oldest people: "Everyday is a gift when you are over 100." *National Geographic* 143 (1): 93–119.

Leed, E. J. 1991. *The mind of the traveler: From Gilgamesh to global tourism*. New York: Basic Books.

Lefebvre, H. 1991. *The production of space*. Malden, MA: Blackwell.

Leitner, G. W. 1889. *The Hunza and Nagyr handbook*. London: Woking Orientalist Institute.

Leopold, A. 1987. *A Sand County almanac: And sketches here and there*. 1949. Reprint, New York: Oxford University Press.

Lewis, M. 2003. *Inventing global ecology: Tracking the biodiversity ideal in India, 1945–1997*. New York: Orient Longman.

Li, T. 2000. Articulating indigenous identity in Indonesia: Resource politics and the tribal slot. *Comparative Studies in Society and History* 42 (1): 149–79.

Lin, H.-T. 2009. The tributary system in China's historical imagination: China and Hunza, ca. 1760–1960. *Journal of the Royal Asiatic Society* 19 (4): 489–507.

Lipman, J. 1997. *Familiar strangers: A history of Muslims in northwest China*. Seattle: University of Washington Press.

Lorimer, E. 1939. *Language hunting in the Karakoram*. London: George Allen and Unwin.

Ludden, D. 2000. Agrarian histories and grassroots developments in South Asia. In *Agrarian environments: Resources, representations and rule in India*, edited by K. Sivaramakrishnan and A. Agarwal, 251–64. Durham: Duke University Press.

MacCannell, D. 1976. *The tourist: A new theory of the leisure class*. New York: Schocken.

Mangan, J. A. 1986. *The game ethic and imperialism: Aspects of the diffusion of the ideal*. London: Penguin Books.

Marsden, M. 2008. Muslim cosmopolitanism? Transnational life in northern Pakistan. *Journal of Asian Studies* 67 (1): 213–47.

Marsh, B. D. 2009. Ramparts of empire: India's north-west frontier and British imperialism, 1919–1947. PhD diss., University of Texas, Austin.

Mauss, M. 1979. *Seasonal variations of the Eskimo: A study in social morphology*. London: Routledge.

Mbembe, A. 2003. Necropolitics. *Public Culture* 15 (1): 11–40.

McCarrison, R. 1937. Adventures in research: The Lloyd Roberts Lectures given before the Medical Society of London. *Transactions of the Medical Society of London* 60:46–71.

———. 1945. *Studies in deficiency diseases*. London: Henry Frowde and Hoddler and Stoughton.

———. 1961. *Nutrition and health: Being the Cantor Lectures delivered before the Royal Society of Arts, 1936, together with two earlier essays*. London: Faber and Faber.

McFarlane, R. 2003. *Mountain of the mind: A history of a fascination*. London: Vintage.

McGrane, B. 1989. *Beyond anthropology: Society and the other*. New York: Columbia University Press.

McWilliams, A. 2007. Harbouring traditions in East Timor: Marginality in a lowland entrepot. *Modern Asian Studies* 41 (6): 1113–43.

Meeker, M. 1980. The twilight of a South Asian heroic age: A rereading of Barth's study of Swat. *Man* 15 (4): 682–701.

Michaud, J. 2010. Zomia and beyond. In Zomia and beyond, edited by J. Michaud. Special issue, *Journal of Global History* 5 (2): 289–312.

Miller, D. 1995. *On nationality*. Oxford: Clarendon.

Mills, C. W. 2000. *The sociological imagination*. 1959. Reprint, New York: Oxford University Press.

Mock, J. 1998. The discursive construction of reality in the Wakhi community of northern Pakistan. PhD diss., University of California, Berkeley.

Molebatsi, C. 2002. The Remote Areas Development Programme and the integration of Basarwa into the mainstream of Botswana. *Pula: Botswana Journal of African Studies* 16 (2): 123–34.

Molina, A., and Á. Esteban. 2006. Tourism brochures: Usefulness and image. *Annals of Tourism Research* 33 (4): 1036–56.

Montgomerie, T. G. 1871. Reports of the mirza's exploration from Cabul to Kashghar. *Journal of the Royal Geographical Society* 41:32–93.

———. 1872. A havildar's journey through Chitral to Faizabad in 1870. *Journal of the Royal Geographical Society* 42:180–201.

Moorcroft, W., and G. Trebeck. 1841. *Travels in the Himalayan provinces of Hindustan and the Panjab; in Ladakh and Kashmir; in Peshawar, Kabul, Kunduz, and Bokhara, 1819–1825*. Edited by H. Wilson. London: John Murray.

Moore, S. D., J. Kosek, and A. Pandian. 2003. *Race, nature and the politics of difference*. Durham: Duke University Press.

Morris-Suzuki, T. 1999. Lines in the snow: Imagining the Russo-Japanese frontier. *Pacific Affairs* 72 (1): 57–77.

Myers, J. 1942. Nomadism. *Man* 42:16–17.

Nash, R. 1967. *Wilderness and the American mind*. New Haven: Yale University Press.

Nazeem Khan, M. 2001. *The autobiography of Sir Mohomed Nazim Khan, KC*. Translated by SherBaz Burcha. Gilgit: Honey-sara.

Netting, R. 1981. *Balancing on an Alp: Ecological changes and continuity in a Swiss mountain community*. Cambridge: Cambridge University Press.

———. 1990. Links and boundaries: Reconsidering the Alpine village as ecosystem. In *The ecosystem approach in anthropology*, edited by E. F. Moran, 229–46. Ann Arbor: University of Michigan Press.

Neuhaus, T. 2012. *Tibet in the Western imagination*. New York. Palgrave Macmillan.

Neumann, R. 2002. *Imposing wilderness: Struggles over livelihoods and nature preservation in Africa*. Berkeley: University of California Press.

Oelschlaeger, M. 1991. *The idea of wilderness: From prehistory to the age of ecology*. New Haven: Yale University Press.

Pagenstecher, C. 2001. The construction of the tourist gaze: How industrial was post-war German tourism? In Construction et affirmation d'une industrie touristique au XIXè et au XXè siècle: Technologie, politique et économie—Perspectives internationales / Construction and strengthening of a tourist industry in the 19th and 20th century: Technology, politics and economy—International perspectives. Colloque international préparatoire au XIIIème Congrès international d'histoire économique, January 17–20, Sion, Switzerland.

Pandian, A. 2008. *Crooked stalk: Cultivating virtue in South India*. Durham: Duke University Press.

Paracha, Nadeem. 2013. The "swinging seventies" in Pakistan: An urban history. *Daily Dawn*, August 22. http://www.dawn.com/news/1037584/the-swinging-seventies-in-pakistan-an-urban-history (accessed July 7, 2014).

Parkes, P. 1987. Livestock symbolism and pastoral ideology among the Kafirs of the Hindu Kush. *Man* 22:637–70.

———. 2000. Enclaved knowledge: Indigent and indignant representations of the environmental management and development among the Kalasha of Pakistan. In *Indigenous environmental knowledge and its transformations: A critical anthropological perspective*, edited by R. Ellen, P. Parkes, and A. Biker, 253–91. Amsterdam. Hardwood Academic.

Pearce, P., and G. Moccardo. 1986. The concept of authenticity in tourist experiences. *Journal of Sociology* 22 (1): 121–32.

Piot, C. 1999. *Remotely global: Village modernity in West Africa*. Chicago: University of Chicago Press.

Pratt, M. L. 1993. *Imperial eyes: Travel writing and transculturation*. London: Routledge.

Raffles, H. 2002. Intimate knowledge. *International Social Science Journal* 54 (173): 325–35.

Raman, A. 2004. Of rivers and human rights: The Northern Areas, Pakistan's forgotten colony in Jammu and Kashmir. *International Journal on Minority and Group's Rights* 11:187–228.

Ramusack, B. 2004. *Indian princes and their states*. Cambridge: Cambridge University Press.

Rawlinson, H. 1872. Monograph on Oxus. *Journal of the Royal Geographical Society* 42:482–513.

Redford, K. 1991. The ecologically noble savage. *Cultural Survival Quarterly* 15 (1): 46–48.

Regar, G. 2002. The price histories of some imported goods on independent Delos. In *The ancient economy*, edited by Walter Scheidel and Sitta Von Reden, 133–54. New York: Routledge.

Rizvi, M. 1971. *The frontiers of Pakistan: A study of frontier problems in Pakistan foreign policy*. Karachi: National Publishing House.

Robb, P. 1997. The colonial state and the construction of Indian identity: An example on the northeast frontier in the 1880s. *Modern Asian Studies* 31 (2): 245–83.

Roberts, L., T. Gordon, C. Trotter, F. Younghusband, and G. Curzon. 1896. The Pamirs and the source of the Oxus: Discussion. *Geographical Journal* 8 (3): 260–64.

Robertson, G. S. 1896. *The Kafirs of Hindukush*. London: Lawrence and Bullen.

Rodale, J. 1948. *The healthy Hunzas*. Emmaus: Rodale.

Romano, T. 2002. *Making medicine scientific: John Burdon Sanderson and the culture of Victorian science*. Baltimore: Johns Hopkins University Press.

Salzman, C. P. 2004. *Pastoralists: Equality, hierarchy, and the state*. Boulder: Westview.

Schaller, G. B. 1988. *Stones of silence: Journeys in the Himalaya*. New York: Viking.

Schneider, H. 1980. *Livestock and equality in East Africa*. Bloomington: Indiana University Press.

Schomberg, R. 1935. *Between the Oxus and the Indus*. London: Martin Hopkinson.

Scott, J. 1998. *Seeing like a state: How certain schemes to improve the human condition have failed*. New Haven: Yale University Press

———. 2010. *The art of not being governed: An anarchist history of upland Southeast Asia*. New Haven: Yale University Press.

Shahrani, M. N. 1979. The retention of pastoralism among the Kirghiz of the Afghan Pamir. In *Himalayan anthropology: Indo-Tibetan interface*, edited by J. F. Fisher, 233–50. The Hague: Mouton.

———. 2002. *The Kirghiz and Wakhi of Afghanistan: Adaptation to closed frontiers and war*. Seattle: University of Washington Press.

Shaikh, F. 2009. *Making sense of Pakistan*. New York: Columbia University Press.

Shaw, R. 1871. *Visits to High Tartary, Yarkand, and Kashghar (formerly Chinese Turkestan), and return journey over the Karakoram Pass*. London: John Murray.

Shimshal Environmental Education Board and Hideki Yamauchi. 1999. The Symposium of "Shimshal Environmental Education Program": In search of true consciousness of the nature and human-being, April 1, Gilgit.

Shimshal Nature Trust. 1994–2009. *Fifteen year vision and management plan*. http://www.snt.org.pk/ (accessed February 13, 2014).

Shneiderman, S. 2010. Are the Central Himalayas in Zomia? Some scholarly and political considerations across time and space. In Zomia and beyond, edited by J. Michaud. Special issue, *Journal of Global History*, 5 (2), 289–312.

Shor, J. B. 1955. *After you, Marco Polo*. New York: McGraw-Hill.

Shor, J., and F. Shor. 1953. At world's end in Hunza. *National Geographic* 117: 485–518.

Sidky, H. 1994. Shamans and mountain spirits in Hunza. *Asian Folklore Studies* 53:67–96.

———. 1996. *Irrigation and state formation in Hunza: The anthropology of a hydraulic kingdom*. Lanham, MD: University Press of America.

Skaria, A. 2001. *Hybrid histories: Forests, frontiers and wildness in western India*. Studies in social ecology and environmental history. Oxford: Oxford University Press.

Smith, V., ed. 1977. Introduction to *Hosts and guests: The anthropology of tourism*, 1–20. Philadelphia: University of Pennsylvania Press.

Sokefeld, M. 2005. From colonialism to postcolonialism: Changing modes of domination in the northern areas of Pakistan. *Journal of Asian Studies* 64 (4): 939–73.

Spooner, B. 1973. *The cultural ecology of pastoral nomads.* Reading: Addison-Wesley.

Stafford, R. 1989. *Scientist of the empire: Sir Roderick Murchison, scientific exploration and Victorian imperialism.* Cambridge: Cambridge University Press.

Steinberg, J. 2006. An anatomy of the transnation: The globalization of the Ismaili Muslim community. PhD diss., University of Pennsylvania, Philadelphia.

Stellrecht, I. 2006. Passage to Hunza: Route nets and political process in a mountain state. In *Karakoram in transition: Culture, development and ecology in the Hunza Valley,* edited by H. Kreutzmann, 191–216. Oxford: Oxford University Press.

Stephens, I. 1955. *Horned moon.* Bloomington: Indiana University Press.

Stocking, G. 1968. *Race culture and evolution: Essays in the history of anthropology.* New York: Free Press.

Sturgeon, J. 2005. *Border landscapes: The politics of Akha land use in China and Thailand.* Seattle: University of Washington Press.

Talbott, I. 2005. *Pakistan: A modern history.* New York. Palgrave Macmillan.

Telegraph (Calcutta). 2012. State ready to revise remote list by special correspondent, January 21. http://www.telegraphindia.com/1120121/jsp/bengal/story_15033818.jsp (accessed July 7, 2014).

Tobe, J. 1960. *Hunza: Adventure in a land of paradise.* St. Catherines, Ontario: Provoker.

Todd, H. 1951. The Sinkiang-Hunza frontier. *Journal of the Royal Central Asian Society* 38 (1): 73–81.

Toor, S. 2005. A national culture for Pakistan: The political economy of a debate. *Inter-Asia Cultural Studies* 6 (3): 318–40.

Trautmann, T. 1997. *Aryans and British India.* Berkeley: University of California Press.

Trek to High Mountains. n.d. http://www.trektohighmountains.com/about.php (accessed February 13, 2008).

Trench, F. 1869. *The Russo-Indian question.* London: Macmillan.

Trevelyan, R. 1987. *The golden oriole: A 200-year history of an English family in India.* New York: Viking.

Trotter, H. 1899. The proceedings of the Pamir Boundary Commission. *Geographical Journal* 13 (4): 442–48.

Tsing, A. 1993. *In the realm of the diamond queen: Marginality in an out-of-the-way place.* Princeton: Princeton University Press.

———. 1999. Becoming a tribal elder and other green development fantasies. In *Transforming the Indonesian uplands: Marginality, power and production,* edited by T. M. Li, 159–202. London: Berg.

———. 2004. *Friction: An ethnography of global connections.* Princeton: Princeton University Press.

Turner, T. 1995. Representation, collaboration and mediation in contemporary ethnographic and indigenous media. *Anthropology Review* 11 (2): 102–6.

Urry, J. 1990. *The tourist gaze: Leisure and travel in contemporary societies*. London: Sage.

van Schendel, W. 2002. Geographies of knowing, geographies of ignorance: Jumping scale in Southeast Asia. *Environment and planning D: Society and space* 20 (6): 647–68.

Veniukof, M., and J. Mitchell. 1866. The Pamir and the sources of Amu-Daria. *Journal of the Royal Geographical Society of London* 36:248–63.

Verkaaik, O. 2004. *Migrants and militants: Fun and urban violence in Pakistan*. Princeton: Princeton University Press.

Vigne, G. 1842. *Travels in Kashmir, Ladak, Iskardo, the countries adjoining the mountain-course of the Indus, and the Himalaya, north of the Panjab*. London: Henry Colburn.

Visser-hooft, J. 1925. *Among the Karakoram glaciers in 1925*. London: Edward Arnold.

Wallace, A. R. 2000. *The Malay Archipelago*. North Clarendon, VT: Periplus.

Wallerstein, I. 1974. *The modern world-system*, vol. 1, *Capitalist agriculture and the origins of the European world-economy in the sixteenth century*. New York: Academic.

Watson, G. 2002. Representation of Central Asian ethnicities in British literature. *Asian Ethnicity* 3 (2): 137–51.

West, P. 2008. Tourism as science and science as tourism. *Current Anthropology* 49 (4): 597–626.

Whorton, J. 2000. Civilisation and the colon: Constipation as the "disease of diseases." *British Medical Journal* 321:1588.

Williams, R. 1973. *The country and the city*. New York: Oxford University Press.

Wilson, S. R. 2002. *A look at Hunza culture*. Islamabad: National Institute of Pakistan Studies / Summer Institutes of Languages.

Wittfogel, K. 1957. *Oriental despotism*. New Haven: Yale University Press.

Wolf, E. 1982. *Europe and the people without history*. Berkeley: University of California Press.

Wood, J. 1841. *Narrative of a journey to the source of the river Oxus*. London: John Murray.

Woodman, D. 1969. *Himalayan frontiers: A political review of British, Chinese, Indian and Russian rivalries*. London: Barrie and Rockliff, Cresset.

Worster, D. 1979. *Dust Bowl: The southern plain in the 1930s*. Oxford: Oxford University Press.

———. 2006. *Nature's economy: A history of ecological ideas*. Cambridge: Cambridge University Press.

Wrench, G. T. 1938. *The wheel of health: The source of long life and health among the Hunza*. New York: Dover.

Younghusband, F. 1896. *The heart of a continent—A narrative of travels in Manchuria, across the Gobi Desert, the Pamirs, the Himalayas and Chitral*. London: John Murray.

———. 1924. *Wonders of the Himalayas*. New York: E. P. Dutton.

Yule, H. 1872. Paper connected with the Upper Oxus region. *Journal of the Royal Geographical Society of London* 42:438–81.

Zurick, D. 1992. Adventure travel and sustainable tourism in the peripheral economy of Nepal. *Annals of the Association of American Geographers* 82 (4): 608–28.

India Office Records

India Office Records: Crown Representative: Indian States Residencies Records, c. 1789–1947. (Kashmir Residence Office and Jammu & Kashmir State Files.) British Library, (London).

R/2 /1079/246. 1886. News that Safdar Ali Khan murdered by a gun shot his father Raja Ghazan Khan of Hunza and with the advice of the Wazirs and officials succeeded the chiefship of Hunza.

R/2/1079/245. 1888. Maharaja of Kashmir's intimation of his wish to withdraw the garrison of his own troops from Chaprot: Disturbances in Gilgit in the year 1888.

R/2/1061/1–8. 1889. Kashmir Residency records on subject including settlement operations, Tibet affairs, matter relating to Gilgit Agency and Hunza Nagar tribes.

R/2/1079/248. 1890. Correspondence with the British agent, Gilgit.

R/2/1062/16. 1891. Kashmir residency records on subject including settlement operations, Tibet affairs, matter relating to Gilgit Agency and Hunza Nagar tribes (annual hostages to Kashmir Durbar).

R/2/1079/251. 1894. Hunza and Nagar subsidies: 1. Proposed renewal of the subsidies paid by the British Government and the Kashmir state to the Hunza and Nagar States prior to 1891; 2. Sanction of the Government of India, with effect from the 1st April 1895, to the grant of an annual allowance of Rs. 2,000 to the States of Hunza and Nagar.

R/2/1079/256. 1904. Hunza claims to the Raskam valley.

R/2/1080/257. 1905. Allotment and distribution of land among the people of Hunza; Hunza relation with China; Provision of land for Hunza people in Gilgit.

R/2/1082/265. 1914. Hunza's rights in Raskam.

R/2/1082/266/4C I. 1915. Mir of Hunza's rights in Raskam.

R/2/1082/267/4C II. 1915. Hunza's rights to grazing fees in Taghdumbash Pamir.

R/2/1086/312. 1923. Hunza rights in the Raskam Valley.

R/2/1067/93. 1927. 1. Political control of the independent territory lying between the north west frontier province and the Gilgit Agency; 2. Political and administrative status of the Gilgit Agency and the Gilgit Wazarat; 3. Boundary between Gilgit Agency and Chitral.

L/P&S/12/3292. 1933–40. Hunza: situation in Raskam; Darwaza incident.

R/2/1085/296. 1937. 1. Compensation to the Mir of Hunza for the loss of his rights to Chinese Turkestan; 2. Grant of a Jagir to the Mir of Hunza in the Gilgit sub-division and increase his subsidy.

R/2/1086/299. 1938. 1. Accession of Indian States to Federation; 2. Position of Hunza, Nagir, Yasin, Koh Ghizer, Ishkuman, Chalas etc: in the event of Kashmir acceding to Federation (View that these States and districts are not an integral part of Kashmir).

R/2 /1086/301. 1940. Report on the visit of the resident in Kashmir to Hunza to install Mir Mohammad Ghazan Khan as the Mir of Hunza.

R/2/1070/137. 1942. Area of Jammu and Kashmir state census, 1941.

R/ 2/1071/162. 1945. Redistribution of Waziri and other allowances in Hunza.

R/1/1/359. 1908. Confidential note on Kashmir affairs for 1907–1908, by Major Sir Francis Younghusband, K. C. I. E., resident in Kashmir.

Index

Note: Page numbers in italic type indicate illustrations.